Opportunities for Including the Information Environment in U.S. Marine Corps Wargames

CHRISTOPHER PAUL, YUNA HUH WONG, ELIZABETH M. BARTELS

Prepared for the United States Marine Corps
Approved for public release; distribution unlimited

RAND NATIONAL DEFENSE RESEARCH INSTITUTE

For more information on this publication, visit www.rand.org/t/RR2997

Library of Congress Cataloging-in-Publication Data is available for this publication.
ISBN: 978-1-9774-0468-8

Published by the RAND Corporation, Santa Monica, Calif.

Preface

This report documents the results from a RAND study that explored considerations related to the way in which the information environment could be better included in U.S. Marine Corps wargames. Findings should be of interest to three main groups of stakeholders: First, members of the community of practice for information-related capabilities and operations in the information environment, who should be interested in approaches to wargaming their capabilities and ensuring that their contributions are reflected in wargames; second, members of the defense wargaming community of interest, who should find the models and taxonomies useful for designing and structuring games and procuring gaming and simulation tools that include the information environment; and third, players and participants across the Marine Corps and U.S. Department of Defense who want wargames that consider all the major factors that contribute to operational outcomes, including information.

This research was sponsored by the Marine Corps Information Operations Center and conducted within the Navy and Marine Forces Center of the RAND National Defense Research Institute, a federally funded research and development center sponsored by the Office of the Secretary of Defense, the Joint Staff, the Unified Combatant Commands, the Navy, the Marine Corps, the defense agencies, and the defense intelligence enterprise.

For more information on the RAND Navy and Marine Forces Center, see www.rand.org/nsrd/nmf or contact the director (contact information is provided on the webpage).

Contents

Figures and Tables

Figures

Tables

Summary

This report presents recommendations to improve the wargaming of operations in the information environment (OIE) and to consider the information environment (IE) in wargames in general. The central question of this research was "What solutions or alternative structures can be included in wargames to adequately portray effects in and through the IE and allow adjudication to consider factors other than inflicting casualties?"

To conduct this research, we reviewed recent joint and Marine Corps doctrine and concepts and spoke with a range of stakeholders to identify requirements for including OIE and the IE in wargames. A review of existing tools and practices revealed shortcomings in wargaming, which all too often neglects OIE and lacks the tools to adequately represent the IE. To identify potential solutions, we explored military theory regarding the mental, morale, and physical spheres of warfare and a range of alternative defeat mechanisms.[1] If future wargames move away from attritional combat models and embrace broader conceptions of objectives and defeat in warfare, they can not only better incorporate OIE but also better represent the physical aspects of warfighting.

> "The information environment is the aggregate of individuals, organizations, and systems that collect, process, disseminate, or act on information.
>
> Joint Publication (JP) 3-13, *Information Operations*, Washington, D.C.: U.S. Joint Chiefs of Staff, incorporating change 1, November 20, 2014, p. 1-1.

> "The information environment comprises and aggregates numerous social, cultural, cognitive, technical, and physical attributes that act upon and impact knowledge, understanding, beliefs, world views, and, ultimately, actions of an individual, group, system, community, or organization.
>
> JP 3-0, *Joint Operations*, Washington, D.C.: U.S. Joint Chiefs of Staff, incorporating change 1, October 22, 2018, pp. IV-1–IV-2.

[1] The concept of mental, moral, and physical spheres comes from the writings of the prolific (and controversial) strategist and British Army officer John Fredrick Charles (J.F.C.) Fuller in the period between World Wars I and II. However, many others have used this construction. When these theorists use the term *moral*, the contemporary meaning of what they denote is much closer to *morale* (having to do with fighting spirit, élan, and will to fight) than with ethics or adherence to principals of right or wrong. Thus, we have substituted *morale* throughout this report.

The IE is Receiving Increased Attention Within the U.S. Department of Defense and Marine Corps, but It Is Not Receiving Commensurate Attention in Wargaming

There is a growing emphasis on the IE within the U.S. Department of Defense (DoD) and the U.S. Marine Corps. The 2018 *Joint Concept for Operating in the Information Environment*, the addition of *information* as a seventh function in joint doctrine, the establishment of a deputy commandant within the Marine Corps for information, and the creation of the Marine Expeditionary Force Information Groups all point to the rising importance attached to the role of information in warfighting.

Increased DoD attention to the IE has not happened in a vacuum. Rather, this shift has occurred alongside growing adversary emphasis on information. Information has become increasingly important in the military thinking of near-peer competitors, such as Russia and China, with Russia—in particular—demonstrating a commitment to developing the capabilities needed to compete in the IE.

Wargaming has also enjoyed new prominence in defense policymaking and offers an important tool to explore potential future conflicts and shape strategy in a low-risk environment. Incorporating the IE into wargames in a meaningful way should be important. Unfortunately, the IE is not well represented in many DoD games, possibly due to a lack of expertise or experience among those who design and commission such games. This failure to include the IE in games can result in an overreliance on attrition-based approaches to adjudication. Too often, wargames ignore the IE or focus too narrowly on specific issues—such as situational awareness, the fog of war, and command and control (C2)—and not on OIE or effects in and through the IE. As a result, many DoD games are not able to adequately represent the role of the IE in a conflict, which can lead to inaccurate analysis, negative learning, and potentially detrimental real-world outcomes.

Requirements for Wargaming Operations in the Information Environment

We identified six key types of information that can be relevant to military operations and thus might need to be represented in wargames (depending on the game's objectives and level of abstraction):

1. situational awareness and situational understanding (including battle damage assessment)
2. C2, including communication
3. C2 warfare (C2W) and other factors that degrade situational awareness or C2
4. information or aspects of the IE that can cause subordinates to behave in ways contrary to the commander's orders or preferences

5. efforts to protect against contrary subordinate behavior or to encourage such behaviors among adversary forces (information for effect)
6. factors in and through the IE that affect the perceptions or behaviors of relevant actors other than adversaries.

Building on these general requirements, we identified requirements specific to the Marine Corps. Various stakeholders indicated that there was a need for greater depth in IE scenarios, a need for more feedback from and about the IE, and a need to include the cognitive aspects of the IE in wargames. Wargames also need to incorporate the various information-related capabilities (IRCs) employed by the Marine Corps, including intelligence, C2, cyber operations, military information support operations, civil-military operations, space operations, electromagnetic spectrum operations, communication strategy and operations, operation security, information assurance, physical security, key leader engagement, defense support to public diplomacy, physical attack, military deception, and signature management.

Considering the general requirements for including the IE in wargaming, specific requirements drawn from discussions with stakeholders, and the various capabilities that the Marine Corps considers explicitly tied to OIE led us to compile the following summary list of requirements:

- game mechanics that accommodate all six categories of information
- games that include IE-related scenarios
- games that capture OIE with greater depth and realism
- mechanisms to incorporate effects of OIE on noncombatant populations
- mechanisms to show effects of noncombatant populations on the IE and the operating environment
- mechanisms to capture the impact of other actors in the IE (e.g., coalition partners)
- games that provide feedback to players related to the IE and OIE
- ability to include assessment of OIE in games
- ability to provide in-game situational awareness about the IE (rather than just situational awareness for the spatial domains)
- games that include the effects of the IE on combat
- ability to perform signature management
- games that include deception or surprise
- games that include cyber or other technical capabilities
- games that include electronic warfare
- games that include various forms of influence—on adversaries or other relevant actors.

Table S.1 presents sample solutions to challenges associated with the requirement to accommodate all six categories of information. The solutions in Table S.1 follow a "crawl, walk, run" progression, with solutions on the right having greater fidelity but also being more difficult or complex to implement.

Table S.1
Crawl, Walk, and Run Solutions to Incorporate Six Categories of Information into Wargames

Information Category	Crawl Solution	Walk Solution	Run Solution
1. Situational awareness and situational understanding	Double-blind games with three maps (one "ground truth" held by the control cell and one map each for red and blue); rules for detecting, sensing, and updating; and manual adjudication	Digitized double-blind control for displaying what players should see	Sophisticated models that track not only locations but also, e.g., unit-sensing or -detecting radii, discrimination capabilities, and different levels of detection or identification
2. C2, including communication	Subject-matter expert (SME)–imposed delays of updates to maps, delays or degradation of messages, or no new orders to units outside the reach of communication systems	Systematic structure, rules, or tables dictating time required to see effects or receive messages, based on battlefield and IE conditions	Software supported by sophisticated models for, e.g., update delays, lags or disruptions in communications
	Physical separation of players at different locations or echelons, with intentional time delays in communication between them, as appropriate	Software that imposes rule- or table-based delays in updates to information	
3. C2W and other factors that degrade situational awareness or C2	Crawl solutions for categories 1 and 2, plus additional rules or SME adjudication of C2W and other IRC actions to increase fog of war or delays or to affect perceptions	Walk solutions for categories 1 and 2, plus additional software rules for C2W actions, their adjudication, and their effects	Run solutions for categories 1 and 2, plus additional complex software that adjudicates specific networks and their connections and presents both the direct impact of various C2W actions and their cumulative or interactive effects
	Log of physical environment and IE characteristics so that players who are deceived or outflanked can see (and accept) how game play actually unfolded	Detailed software-based log of physical environment and IE characteristics so that players who are deceived or outflanked can see how game play actually unfolded	

Table S.1—Continued

Information Category	Crawl Solution	Walk Solution	Run Solution
4. Information that causes subordinates to behave in ways contrary to the commander's orders or preferences	Manual rules, guidelines, or tables for the behavior of units outside the reach of command, leadership or command tests (and situational modifiers), and the quality and characteristics of subordinates	Software-based tracking of circumstances leading to crawl-level situational modifiers and digital tracking of leader characteristics or states and digital tests for compliance	Sophisticated simulation, with a full simulation needed for multiagent models and complex decision rules
	SME injection or adjudication when circumstances cause subordinates to behave differently		SME input conveying deep understanding of human dynamics, the cognitive and cultural biases of relevant decisionmakers, and other behavioral factors
5. Information for effect	Crawl solution for category 4, plus rules or adjudication for efforts to affect or exploit circumstances	Walk solution for category 4, plus digital tracking or incorporation of effects based on relevant circumstances	Run solutions for categories 1, 2, and 4, when unit actions of reflect, e.g., characteristics and proclivities, awareness and perception of the situation, and last orders received (and how long ago they were received)
6. Factors that affect relevant actors other than adversaries	SMEs in control cell loosely track relevant actors, their changing sentiments, and their actions in response to OIE and other actions or events	Mechanism to track all relevant actors and their states and conditions; systematic structure or rules for how game actions or events can affect those states and conditions; systematic structure or rules for the resulting consequences of relevant actors' actions in the game	Sophisticated software-based modeling, including advanced social science, narrative, human dynamics, and cognitive and cultural biases

Military Theory Suggests a Better Approach to Gaming

Military theory acknowledges the importance of three spheres in warfare: mental, morale, and physical. Existing approaches to military analysis tend to focus primarily on the physical, but the other two spheres are critical not only in accurately reproducing OIE in a wargame context but also in accurately capturing the nature of conflict. For wargames to meaningfully reflect the impact of OIE, they must consider all three spheres and allow game actions other than physical movement and physical combat to have other-than-physical effects. Fortunately, the same three-sphere adjudication mechanics that would allow OIE to be meaningfully represented in wargames will also capture the impact of physical combat actions even better than current attrition-based models by allowing destruction to have an explicit—rather than implicit—impact on the mental and morale spheres.

Detailed consideration of how defeat works across all three spheres also suggests a much broader range of defeat mechanisms than the traditional focus on attrition. We identified the following mechanisms beyond simple destruction or attrition: annihilation, dislocation, circumvention, exhaustion, disruption, isolation, surrounding, disintegration, preemption, forestalling, and compellence/impellence/deterrence. Because defeat is the primary objective of military campaigns, it makes sense that the implicit objective given to players in operational wargames is to defeat adversaries. If OIE can meaningfully affect defeat outcomes in games, then players will be more likely to employ such capabilities.

Table S.2 describes sample game mechanics that could allow the application of the various different defeat mechanisms and their effects in all three spheres. Such game mechanics would allow players to employ physical capabilities and IRCs in pursuit of game objectives, meeting the requirements for wargaming OIE while still allowing physical combat to play a central role in warfare.

Table S.2
Sample Game Mechanics to Adjudicate Defeat Mechanisms in the Mental, Morale, and Physical Spheres

Mechanism	Mental Sphere: Perception, Judgment, Decisionmaking	Morale Sphere: Morale, Will, Leadership	Physical Sphere: Capability
Destruction, attrition, or annihilation	For subordinate units, set probabilistic breakpoints that consider a range of mental, morale, and physical factors.	As causalities are adjudicated from force-on-force engagements, the measure of a unit's will to fight is reduced for the remaining forces.	Design meaningful degradation steps for each unit type. Consider how unit capabilities decline in response to attrition.
Dislocation and circumvention	Set time requirements to change orders to ensure that units cannot respond artificially quickly to changing conditions.	Establish rules for shock effect and surprise, as well as changes to the breakpoint.	Impose physical consequences for units attacked on flanks or by surprise.
Exhaustion	Forces for nations nearing exhaustion may be more risk-averse, inclined to both avoid unnecessary losses in a losing effort and avoid results that further contribute to domestic war-weariness.	Track national will to fight and factors that reduce it. Units may face declining morale alongside decline in domestic support for the mission.	As national will to fight is exhausted, reinforcements and supplies may be reduced or operational constraints may be imposed.

Table S.2—Continued

Mechanism	Mental Sphere: Perception, Judgment, Decisionmaking	Morale Sphere: Morale, Will, Leadership	Physical Sphere: Capability
Disruption, isolation, and surrounding	Track lines of communication. Units that are not "in command" assume default behavior (e.g., following last available order, responding in an adjudicator-determined manner to conditions, doing nothing).	Track lines of communication. Units that are not "in command" receive a will-to-fight penalty.	Track key unit supplies and expenditures. Units low on supplies are physically hindered (e.g., no food) or cannot act (e.g., no ammunition).
Disintegration	Subordinate units may be modeled as more likely to fail to respond to orders.	Low will to fight can result in unit breakpoint behavior, under which orders cannot be carried out.	
Preemption	Natural player confusion can be heightened by restricting the time available to plan moves in response to a surprise attack. Subordinate units may be modeled to be more likely to fail to respond to orders.	Surprise attacks create a will-to-fight penalty.	
Forestalling	Stalled forces cannot accept new orders until they successfully disengage.	Stalled forces take a will-to-fight penalty.	Unit cannot physically move until winning an engagement. (These mechanics are known as *zones of control* in commercial games.)
Compellence/ impellence/ deterrence	Simulated higher-level commanders' orders can terminate the war or change objectives based on outcomes.		

Conclusions and Recommendations

The IE and OIE are increasing points of emphasis for both DoD and the Marine Corps, as well as U.S. adversaries. Wargames can be incredibly valuable, but their value is diminished if they do not consider the IE. If the IE is important in operations, it should be important in wargames, yet the IE, OIE, and IRCs are all frequently underrepresented.

Drawing on current doctrine, military theory, and discussions with stakeholders, we offer recommendations targeting three key communities: the sponsors of wargames, designers of wargames, and those who procure new tools and recruit personnel to support wargaming.

Everyone involved in wargaming should acknowledge the role of information in operations and seek to better represent the relevant aspects of the IE in games. Furthermore, stakeholders should recognize the potential role for OIE across the spectrum of conflict. The effects of OIE do not remain in the IE; such operations have consequences in and across the spatial domains. OIE are not just useful in affecting noncombatants; these operations also have a role in combat against peer or near-peer adversaries. And OIE are not just relevant in phase 0 or during competition; these tactics are useful in all phases and at all intensities of conflict. Game accordingly.

Recommendations for Wargame Sponsors

Wargame sponsors have the greatest potential to influence trends in wargaming by ensuring that games serve a broader purpose of preparing forces for realistic operational scenarios, which will inevitably be influenced by the IE. Sponsors are also best positioned to ensure that a game's content captures a range of information types and defeat mechanisms in service of larger strategic and operational goals. The following recommendations can help guide sponsors as they plan for games, evaluate designs, and oversee play and follow-on analysis:

- Explicitly emphasize the role of the IE and its relevance to the game's purpose.
- Move from critiquing the absence of OIE to demanding their inclusion in games. Require that games move away from attrition-centric models to models that more fully consider the mental and morale spheres of conflict in ways that allow players to make meaningful decisions and that project plausible outcomes in meaningful ways.
- Explicitly require game designers to include OIE and make that requirement clear during scoping conversations.
- Explicitly require that OIE remain central to game play, and ensure that there is a procedure for documenting these operations and their effects in post-game analysis to enable evaluation and learning by players, other sponsors, and wargame designers.
- If the relevance of the IE is not evident at each stage of a game's design and execution, engage with designers to understand how OIE are being included.
- Be prepared to support additional recruitment efforts to bring in nontraditional players and subject-matter experts who can enhance the game's ability to adequately capture the IE. To further improve the fidelity of game play, consider sponsoring the development of OIE-related educational materials for players.

Recommendations for Wargame Designers

Wargame designers are best positioned to identify options for incorporating the IE into games, working alongside sponsors to ensure that how a game does so aligns with the sponsor's objectives. The following recommendations will help designers consider the IE from the earliest stages of planning:

- Identify which of the six types of information are important to the overall purpose or objectives of the game and find ways to incorporate these interactions.
- When developing new systems, start from the notion that the mental, morale, and physical aspects of a conflict all need to be represented. Because the tools for representing physical aspects are more established, holistically representing the conflict will require devoting substantial time to the development of mental and morale approaches.
- Remember that representing more spheres of conflict will add more elements to the game and thus more complexity. Think hard about how to balance complexity and abstraction to most usefully characterize the key challenges within the game.
- Remember that IE expertise may not reside among the "usual suspects" in the wargaming community. Be prepared to bring in new SMEs to help evaluate, test, and refine new game mechanics and underlying models.
- Remember that all cells should have players who are familiar with OIE. Recruit accordingly, and be prepared to recruit from new communities or to build educational content into the game.
- Ensure that post-game analysis considers the aspects of the game's design or play that affected player decisionmaking, particularly with respect to information, and determine why players made the decisions they made.

Recommendations for Those Who Procure and Provide Wargaming Capabilities

Wargame design and execution requires various types of support, including the procurement of game materials, technologies, and expertise. The following recommendations apply to the procurement of software and tools to support game development or play:

- Select tools that can represent the mental and morale spheres and a range of defeat mechanisms, information, and conditions that can affect game play (which may be unknown to players), along with robust models of human dynamics, psychological factors, and information flows.
- Ensure that systems are transparent to allow designers to validate how OIE are incorporated and their effect on game play.
- Check that the weights of specific types of information, operational actions, and noncombatant behaviors can be updated and modified as understanding of OIE matures, and consider how these lessons can be applied to new regions and conflicts.

- Select systems that can represent the types of information needed by involving game designers and sponsors in the evaluation process.

The following recommendations are intended to guide the recruitment of personnel to support game design, testing, and execution or the involvement of SMEs in specific aspects of these tasks:

- Specify requirements for expertise in such areas as qualitative research and social science, evolving OIE practice, and regional populations when writing statements of work, task orders, and position descriptions.
- Representing OIE requires knowledge of relevant operational art, adversary doctrine and practices, and local context. It may not be practical to maintain staff with expertise in these areas at all times, but efforts should be made to recruit supplemental experts to support specific game requirements. Capability developers should form a "stable" of relevant IE and OIE expertise by requesting SME rosters from other U.S. military and government organizations and by developing relationships with outside experts.
- Because the concepts and practices associated with OIE are evolving, OIE expertise is not static. Experts should refresh their knowledge as needed to stay up to date with new practices and evolving contexts.

Acknowledgments

We are indebted to David Grohoski and LtCol Earnest Robinson at the Marine Corps Information Operations Center for their guidance and assistance on this project. We also thank the IRC and information operations (IO) stakeholders who were generous with their time and insights. The gaming community, including Marine Corps wargamers, were also enormously helpful in providing background and insights over the course of our project. We greatly appreciate the opportunity to work with them on such meaningful and important topics.

Our gratitude goes out to the various stakeholders within the wargaming community and in the IO/OIE/IRC community. The terms of our interviews prevent us from acknowledging them by name, but their input was invaluable.

We also thank our colleagues Jon Fujiwara at RAND and Sara Cobb at George Mason University for their contributions to the project's initial framing and preliminary thinking. We extend further thanks to our two reviewers, who provided comments and feedback on the draft as part of RAND's quality assurance process: Colin Clarke and Ed McGrady. We are also grateful to RAND Navy and Marine Forces Center director Paul DeLuca and former associate director Michael Decker for their oversight. Administrative support and citation formatting were helpfully provided by Maria Falvo. The RAND editorial team, including editor Lauren Skrabala and production editor Matthew Byrd, have done amazing work purging errors, improving readability, and otherwise preparing this report for public consumption. Finally, we thank Rick Penn-Kraus, who designed the report's cover.

Abbreviations

AFSIM	Advanced Framework for Simulation, Integration, and Modeling
AWARS	Advanced Warfighting Simulation
C2	command and control
C2W	command-and-control warfare
C4ISR	command, control, communications, computers, intelligence, surveillance, and reconnaissance
DoD	U.S. Department of Defense
IE	information environment
IO	information operations
IRC	information-related capability
ISR	intelligence, surveillance, and reconnaissance
JP	joint publication
JSAF	Joint Semi-Automated Forces
MAGTF	Marine air-ground task force
MEF	Marine expeditionary force
MIG	Marine expeditionary force information group
MILDEC	military deception
MISO	military information support operations
OIE	operations in the information environment
OPSEC	operations security

SIGMAN	signature management
SME	subject-matter expert
SSA	support for strategic analysis
SWIFT	standard wargaming integration and facilitation tools
VAST	versatile assessment simulation tool

CHAPTER ONE

Introduction

Purpose and Scope

This report reviews how operations in the information environment (OIE) are currently represented in wargames and recommends improvements.[1] The U.S. Marine Corps and the joint community view OIE as an increasingly important aspect of warfare. At the same time, the Marine Corps is making sizable investments in new wargaming capabilities. Yet, OIE, the information environment (IE) more broadly, and the various information-related capabilities (IRCs) generally remain underdeveloped and underrepresented in wargames, leaving many opportunities to incorporate IE dynamics more consistently and in a more meaningful way.

The core of the problem appears to hinge on the most common adjudication mechanisms in wargames: Units in games are weakened or removed from play based exclusively on physical damage and attrition-based thresholds. OIE do not typically directly result in casualties. Although many IRCs act as force multipliers, they also contribute effects that cannot be captured within an attrition-focused paradigm. Thus, the search for ways to better include the IE in Marine Corps wargames must include approaches to adjudication that capture both the physical damage and casualties caused by warfighting and the psychological and other effects generated by OIE and IRCs. The central question of this research was as follows:

> What solutions or alternative structures can be included in wargames to adequately portray effects in and through the IE and allow adjudication to consider factors other than inflicting casualties?

---

[1] We adopted Peter Perla's definition of a *wargame* as "a warfare model or simulation whose operation does not involve the activities of actual military forces, and whose sequence of events affects and is, in turn, affected by the decisions made by players representing the opposing sides" (Peter Perla, *The Art of Wargaming: A Guide for Professionals and Hobbyists*, John Curry, ed., History of Wargaming Project, 2011, p. 157).

It is important to note that we use *model* throughout this report to refer to the conceptual representation of reality in games (which may be more or less formalized and is rarely computerized). We use *modeling and simulation* to refer to a computerized, quantitative representation of combat.

Although the U.S. Department of Defense (DoD) employs wargaming across a range of organizations and activities, including the joint and interagency level, the recommendations in this report focus on the wargaming activities that the Marine Corps can most directly influence. This is not to understate the importance of having good OIE representation in Navy–Marine Corps, joint, and interagency games. We are instead hopeful that developing good OIE wargaming practice will allow the Marine Corps to set a precedent that will spur the incorporation of OIE into higher-level games over time.

Approach

We approached the problem by reviewing a variety of existing Marine Corps, other service, and joint wargames. Specifically, we examined their objectives, practice, methods, adjudication techniques, tools, simulations, outcomes, and game-related reports and post-game analyses, when such information was available. This involved observing Marine Corps exercises, the research team's past experience as observers and participants in Marine Corps wargames and exercises, and a review of wargame reports. We also spoke with Marine Corps stakeholders involved with IRC games to get a better sense of their requirements and their experiences with OIE in existing games.[2] We reviewed the broader wargaming literature and practice outside DoD circles and also drew on the team's experiences and contacts in the peace gaming and international defense and national security gaming communities.[3] We also reviewed military theory for principles and mechanisms that might be useful in enhancing or redesigning games to better capture warfighting, both in the spatial domains and in and through the IE. This research was limited by the scope of our inquiry. For example, we did not solicit input from international SMEs or those outside the Marine Corps, and our literature review focused only on material that we expected to be the most relevant.

Based on these discussions and reviews, we formulated recommendations using the familiar framework of "crawl, walk, run" stages of incorporating the IE into deci-

---

2 We spoke with approximately ten subject-matter experts (SMEs) as part of this research. In selecting our interview participants, our goal was not necessarily to identify a representative sample of the larger population but (with guidance from the project sponsor) to capture relevant organizational stakeholder perspectives, such as the Marine Corps wargaming and information operations/OIE communities and relevant IRC specialties).

The semistructured interviews addressed existing capabilities to include the IE and OIE in wargames and requirements related to doing so. Interviews were conducted in person or by phone and lasted between 30 and 60 minutes. RAND's Institutional Review Board reviewed the research plan and draft interview protocol and determined that this was not research involving human subjects.

3 On peace gaming, see, for example, Takeshi Utsumi, "Globally Collaborative Environmental Peace Gaming with Global University System," paper presented at the third International Learning GRID of Excellence Working Group workshop, Berlin, Germany, December 3, 2003.

sionmaking.[4] We emphasize wargaming practices and conceptual mechanisms or abstractions for OIE over specific gaming technologies or tools. This is both because of the importance of incorporating new defeat mechanisms into wargame structures—independent of tools—in improving IE representation in wargames, as well as the relative scarcity of wargaming tools designed to capture important aspects of the IE.

How This Report Is Organized

The remainder of this report proceeds as follows. Chapter Two notes recent changes in DoD and Marine Corps thinking about OIE, arguing that these operations are important to Marine Corps warfighting and that wargames are an important preparation tool for the full range of operations. Thus, it is important to include OIE in wargames. The chapter concludes with a brief discussion of the current status of OIE in wargames.

In Chapter Three, we discuss requirements for wargaming operations in and through the IE. We build from the general to the specific, first identifying six categories of information in warfare that merit consideration in wargames. We then identify requirements for wargaming that are specific to the Marine Corps, drawn from conversations with stakeholders and from lists of relevant capabilities.

Because incorporating IE considerations into wargaming requires a discussion of foundational principles, Chapter Four revisits assumptions about the nature of conflict and the relevance of information. We pay particular attention to the mental, morale, and physical spheres of warfare.[5] Chapter Five identifies a range of alternative defeat mechanisms, their relationship to the three spheres of warfare discussed in Chapter Four, and the implications for wargaming. Chapter Six discusses the concepts of breakpoints and surprise in greater detail. Chapter Seven builds on the previous three chapters to offer specific solutions to address the unmet requirements outlined in Chapter Three.

We present conclusions and recommendations in Chapter Eight, including general conclusions about the importance of the IE and OIE, the associated shortcomings of current wargames, and how wargaming could better integrate the IE. We offer sev-

---

4 See, for example, Christopher Paul, "The Crawl, Walk, Run Progression for the Integration and Conduct of Efforts to Inform, Influence, and Persuade," *IO Sphere*, Fall 2013.

5 *Mental*, *moral*, and *physical* spheres of warfare are described in J. F. C. Fuller's classic writings from the period between World War I and World War II. Many others have used this construction as well. When these theorists reference *moral*, the contemporary meaning is much closer to *morale* (having to do with fighting spirit, élan, and will to fight) than to ethics or adherence to principals of right or wrong. Thus, we have substituted *morale* throughout this report, except where we quote directly from one of these original sources or when discussing ethics and morality.

eral recommendations specifically for wargame sponsors, designers, and those tasked with supporting the acquisition and development of wargaming capabilities.

The report concludes with a series of appendixes that provide additional context for the discussions, conclusions, and recommendations. Appendix A contains a glossary of wargaming terms, and Appendix B defines various information-related terms and concepts used in defense circles to help readers better understand how such terms as *information warfare* and *operations in the information environment* are used in doctrine and in practice. Appendix C presents greater detail on the changes in the Marine Corps related to OIE that were briefly introduced in Chapter Two. Appendix D provides additional detail on wargaming tools that are mentioned throughout this report.

CHAPTER TWO

Operations in the Information Environment and Wargaming

DoD's emphasis on and interest in the IE has waxed and waned over time, but the IE is currently a topic of focus at both the joint and service levels. In this chapter, we describe how interest in the IE has burgeoned in recent years and highlight some of the changes that have occurred as a result of this emphasis. We also discuss why the ability to accurately wargame the IE matters, given these developments.

A glossary of wargaming terminology can be found in Appendix A, and Appendix B provides additional information on defense community terminology and concepts related to information as used in doctrine and practice.

A Growing Emphasis on the IE

DoD's Interest in OIE

The Marine Corps created the position of Deputy Commandant for Information and a series of information groups in the Marine expeditionary forces (MEFs), highlighting the importance of information and the IE to its operations.[1] At the joint level, the IE has emerged as a key area of interest over the past several years, and *information* has been elevated to the status of joint function, joining the six traditional joint functions: command and control (C2), intelligence, fires, movement and maneuver, protection, and sustainment.[2] The 2016 DoD *Strategy for Operations in the Information Environment* led to the *Joint Concept for Operating in the Information Environment* and formal, capabilities-based assessment.[3] DoD senior leaders have repeatedly acknowledged the importance of information in military operations. In the words of then–Chairman of

---

1 Mark Pomerleau, "Marines Look to Dominate in Information Environment," *C4ISRNET*, April 5, 2017a. See Appendix C for more on the evolution of the Marine Corps' focus on the IE and OIE.

2 Joint Publication (JP) 3-0, *Joint Operations*, Washington D.C.: U.S. Joint Chiefs of Staff, incorporating change 1, October 22, 2018.

3 U.S. Department of Defense, *Department of Defense Strategy for Operations in the Information Environment*, Washington, D.C., June 2016; U.S. Joint Chiefs of Staff, *Joint Concept for Operating in the Information Environment (JCOIE)*, Washington, D.C., July 25, 2018.

the Joint Chiefs of Staff Joseph Dunford, information must be baked into joint force thinking "from the ground up."[4] If the IE is important in operations, then it should be important in wargames.

Adversary Interest in the IE

U.S. interest in information has not developed in a vacuum but, rather, alongside a growing emphasis on information among adversaries. Information has become increasingly important to the military thinking of near-peer competitors, such as Russia and China—with Russia, in particular, demonstrating a commitment to developing the capabilities needed to compete in the IE.[5]

Russian information warfare, or "information confrontation," has drawn increased attention in the West since Russia seized Crimea from Ukraine in 2014.[6] Russia's information warfare concept is both broad and integral to preparing for conflict from a position of conventional military inferiority compared with the United States, an approach that may be viewed as "asymmetric" by many U.S. defense professionals.[7] At the international strategic level, Russia seeks to undermine democratic institutions, erode confidence in Western financial markets, aggravate social tensions within Western societies, and encourage conspiracy theories about potential U.S. aggression toward Russia.[8] Russian information warfare today is high-volume and multichannel, projecting political narratives into many countries, using channels ranging from official state outlets to social media and troll farms.[9] Russian military thinking on information is not new. It dates to the 1920s and draws on both Soviet concepts and Russia's experiences in post–Cold War conflicts, such as in Chechnya and Georgia.[10]

Two key ideas in Russian military thinking about information are a belief in the overwhelming importance of nonmilitary means of obtaining favorable outcomes and the view that actions in the information space should be used in peacetime and

---

4 Joseph F. Dunford, Jr., "The Pace of Change," *Joint Force Quarterly*, Vol. 84, 1st Quarter 2017, p. 3.

5 Christopher Paul, Colin P. Clarke, Michael Schwille, Jakub P. Hlávka, Michael A. Brown, Steven S. Davenport, Isaac R. Porche III, and Joel Harding, *Lessons from Others for Future U.S. Army Operations in and Through the Information Environment*, Santa Monica, Calif.: RAND Corporation, RR-1925/1-A, 2018.

6 Keir Giles, *Handbook of Russian Information Warfare*, Rome: NATO Defense College, Research Division, Fellowship Monograph 9, November 2016, pp. 1, 3.

7 See Giles, 2016, p. 3.

8 Todd Helmus, Elizabeth Bodine-Baron, Andrew Radin, Madeline Magnuson, Joshua Mendelsohn, William Marcellino, Andriy Bega, and Zev Winkelman, *Russian Social Media Influence: Understanding Russian Propaganda in Eastern Europe*, Santa Monica, Calif.: RAND Corporation, RR-2237-OSD, 2018, pp. 10–11.

9 Christopher S. Chivvis, *Understanding Russian 'Hybrid Warfare'—and What Can Be Done About It*, testimony before the Armed Services Committee, U.S. House of Representatives, CT-468, March 22, 2017, p. 3; Helmus et al., 2018, p. 12; Paul, Christopher, and Miriam Matthews, *The Russian "Firehose of Falsehood" Propaganda Model: Why it Might Work and Options to Counter It*, Santa Monica, Calif.: RAND Corporation, PE-198-OSD, 2016.

10 Giles, 2016, pp. 17, 33–36.

during open conflict.[11] Chief of the Russian General Staff, General Valery Gerasimov, observed that nonmilitary means outweighed military ones at a ratio of four to one in modern warfare—and Russia views information confrontation as a critical nonmilitary activity.[12] Information warfare and psychological operations are seen as laying the groundwork for victory, and even, at times, able to achieve victory without the need for armed conflict. Russian military thinking acknowledges the primacy of information and considers information warfare as a key lever in its own right, rather than merely a force multiplier in support of conventional combat arms.[13]

Information has also become more of a central theme in Chinese military writing. The People's Liberation Army's "three warfares" concept, approved in 2003, identifies information superiority as a precondition for battlefield supremacy.[14] The three warfares are psychological warfare, media or public opinion warfare, and legal warfare.[15] Psychological warfare is aimed at undermining an enemy's ability to fight and takes both adversary military forces and civilian populations into account. Psychological warfare might target enemy morale, motivation and willingness to fight, leadership, international support, economic strength, or domestic political support. Media or public opinion warfare seeks to build international and domestic support for China's military actions and for Chinese interests in general.[16] It also attempts to mobilize Chinese domestic support to signal resolve.[17] Finally, legal warfare seeks to use international and domestic law to legitimize Chinese policies and to undermine enemy justification and authority.[18]

Informationized conditions and *noncontact warfare* are two other Chinese military concepts that relate to information but in a narrower way than the three warfares. A key concept since 2004, the idea of informationized conditions involves military systems integrated with advanced, networked computer and information systems in ways that offer operational advantages.[19] Prompted by U.S. victory in the 1991 Gulf War,

---

[11] Rand Waltzman, *The Weaponization of Information: The Need for Cognitive Security*, testimony before the Armed Services Committee, Subcommittee on Cybersecurity, U.S. Senate, CT-473, April 27, 2017, p. 3; Giles, 2016, p. 10.

[12] Waltzman, 2017, pp. 3–4.

[13] Giles, 2016, pp. 16–18.

[14] Timothy A. Walton, *China's Three Warfares*, Herndon, Va.: Delex Systems, Special Report 3, January 18, 2012, pp. 4–5.

[15] Peter Mattis, "China's 'Three Warfares' in Perspective," *War on the Rocks*, January 30, 2018.

[16] Walton, 2012, pp. 4–5, 7.

[17] P. Mattis, 2018.

[18] Walton, 2012, p. 9.

[19] Timothy R. Heath, Kristen Gunness, and Cortez A. Cooper III, *The PLA and China's Rejuvenation: National Security and Military Strategies, Deterrence Concepts, and Combat Capabilities*, Santa Monica, Calif.: RAND Corporation, RR-1402-OSD, 2016, pp. ix–x, 35.

the idea behind noncontact warfare is that distance is no longer a barrier in conflict. The new "informationized battlefield" instead requires "noncontact" capabilities, such as long-distance strike, unmanned systems, cyber, and information warfare.[20]

If the IE is important in adversary and competitor thinking, planning, and operations, then it should be important in wargames with scenarios involving those adversaries or competitors.

Why Wargaming Is Important to the Marine Corps

Given the changes summarized so far, including the IE in wargaming is important for a range of reasons. Both DoD and the Marine Corps have renewed their interest in wargaming in recent years. The Marine Corps, in particular, is poised to invest considerable resources into improving service-level wargaming capabilities and increasing the number of annual wargames. Wargaming also remains an important tool for both operators and policymakers, providing an opportunity for learning, exploring, and thinking through the potential consequences of planned or potential operations. It is an established tool in the defense and intelligence communities, and especially useful when strategists are faced with difficult, complex problems and uncertain futures.

Even before the recent revival of a high-level focus on wargaming, both the U.S. Army and U.S. Naval War Colleges published handbooks that sought to better describe and instruct on wargaming practice.[21] Other signs of renewed DoD interested in wargaming include high-level memos on wargaming by the Deputy Secretary of Defense and the Secretary of the Navy, the creation of the Defense Wargaming Alignment Group and a DoD wargame incentive fund, and DoD-sponsored wargaming conferences.[22] The Military Operations Research Society created a wargaming certificate program in 2017 in response to the increased demand for wargamers in the defense community.[23] DoD interest in wargaming has also coincided with renewed

---

[20] Heath et al., 2016, pp. 36–37.

[21] James Markley, *Strategic Wargaming Series Handbook*, Carlisle Barracks, Pa.: Center for Strategic Leadership and Development, U.S. Army War College, July 2015; Shawn Burns, ed., *War Gamers' Handbook: A Guide for Professional War Gamers*, Newport, R.I.: U.S. Naval War College, undated.

[22] Bob Work, Deputy Secretary of Defense, "Wargaming and Innovation," memorandum, Washington, D.C., February 9, 2015; Ray Maybus, Secretary of the Navy, Department of Navy, "Wargaming," memorandum, Washington, D.C., May 5, 2015; Bob Work and Paul Selva, "Revitalizing Wargaming Is Necessary to Be Prepared for Future Wars," *War on the Rocks*, December 8, 2015; Garrett Heath and Oleg Svet, "Better Wargaming Is Helping the US Military Navigate a Turbulent Era," *Defense One*, August 19, 2018; Phillip Pournelle, ed., *MORS Wargaming Special Meeting, October 2016: Final Report*, Alexandria, Va.: Military Operations Research Society, 2017, p. 5; Phillip Pournelle and Holly Deaton, eds., *MORS Wargaming III Special Meeting, 17–19 October 2017: Final Report*, Alexandria, Va.: Military Operations Research Society, April 2018, p. 2.

[23] Military Operations Research Society, "Certificate in Wargaming," registration page, undated a; Military Operations Research Society, email to the authors, October 12, 2018.

interest in wargaming in other countries. The UK Ministry of Defence published its own wargaming handbook in 2017 that—unlike the Army and Navy wargaming handbooks—became doctrine.[24] China has also invested in computerized wargaming over the past decade, beginning with a focus on strategic problems before expanding to interservice wargames and tactical simulations.[25]

Against this backdrop, the Marine Corps expects to further increase its wargaming capabilities to better prepare for future combat. It plans to not only develop and conduct a larger number of wargames but also to boost its investment in technologically sophisticated wargaming methods and tools.[26] Marine Corps Systems Command is currently overseeing the development of a "world-class" wargaming capability that seeks to be data-enabled and analytically rigorous, incorporating computerized modeling and simulation and using in-stride game adjudication.[27] Former Commandant General Robert Neller spoke about his desire for a "Star Trek–like holodeck" for wargaming.[28] And stakeholders within the Marine Corps wargaming community have expressed a desire for more-sophisticated adjudication, visualization, analysis, and knowledge management for future Marine Corps wargaming.

Across the Marine Corps, wargaming informs a range of decisions, and the concept of wargaming encompasses a similarly broad set of activities. Examples include training events and simulations, discussion groups and seminars, planning exercises, reviews of plans, and course-of-action wargaming as part of the Marine Corps Planning Process.[29] Other wargame approaches that place a heavier emphasis on adjudication are currently used by the Marine Corps—for example, matrix games, hex-and-counter games, commercial computer games, commercial board games, and manual games used in combination with modeling, simulation, and analysis.[30] Wargames are used from the tactical to the service level and above and are part of initiatives to inform Marine Corps Title 10 responsibilities to organize, train, and equip the force.[31] Stakeholders involved in wargaming engage in concept development, capabilities develop-

---

[24] UK Ministry of Defence, Development, Concepts, and Doctrine Centre, *Wargaming Handbook*, Swindon, UK, August 2017.

[25] Dean Cheng, "The People's Liberation Army on Wargaming," *War on the Rocks*, February 17, 2015.

[26] Todd South, "Marine Wargaming Center Will Help Plan for Future Combat," *Marine Corps Times*, September 19, 2017.

[27] Program Manager Wargaming Capability, Marine Corps Systems Command, "Wargaming Capabilities," Modern Day Marine 2018 Report to Industry briefing, September 26, 2018, slides 1–2.

[28] James Clark, "The U.S. Marine Commandant Wants a 'Star Trek'–Style Holodeck for Wargaming," *National Interest*, September 30, 2017.

[29] Marine Corps Warfighting Publication 5-10, *Marine Corps Planning Process*, Washington, D.C., 2010, pp. 4-2–4-3.

[30] See Appendix A for more detail on various types of wargames.

[31] U.S. Marine Corps Warfighting Laboratory, Futures Directorate, "Title 10 Wargaming," webpage, undated.

ment, training and education, science and technology development, operational planning, and other activities.

Independent of the widespread and institutionalized presence of wargaming in the Marine Corps, when used properly, the method itself can assist in critical thinking, individual and organizational learning, and trial-and-error exploration of new concepts and warfighting approaches without costing lives or materiel. These qualities will be particularly valuable as the Marine Corps continues its transition from more than a decade of counterinsurgency operations to other forms of warfare that are markedly different but about which the current generation of Marines may have no firsthand knowledge. Wargaming also has the potential to play a particularly important role in preparing for OIE, characterized by evolving concepts and understanding, as well as limited opportunities for experimentation in the real world. Although wargames do not prove or "validate" concepts and approaches, they can be used to teach principles, offer perspectives on what does not work, and create additional insights. They have the potential to highlight gaps and raise awareness of potential consequences in ways that are more meaningful to participants than traditional strategic analyses.[32]

Why Wargaming the IE Is Important to the Marine Corps

Given the important role wargames can play in informing future Marine Corps operations, it is critical that these games represent the IE accurately and effectively. Adjudicated wargames offer the opportunity for Marines to try out actions against opposing players and to face a range of plausible consequences. Games that include the IE have the potential to illustrate a wider range of consequences than those focused on physical effects—from domestic public outcry about avoidable collateral damage to the effects of partner-nation forces' morale on in-game battlefield outcomes. Wargames have the potential to improve Marine Corps understanding of OIE and the role of information in operations, and they offer an opportunity to refine approaches before Marines face a real-life situation with real consequences.

There are undeniable benefits for those working with IRCs or organizations with a specific focus on information. IRC-specific wargames, centered on cyber operations or signature management (SIGMAN), for example, can improve understanding and better prepare Marines. Wargames also offer a potential avenue to practice the functions and relationships outlined in the Marine air-ground task force (MAGTF) concept of employment for IE operations or to explore ways to improve upon initial MEF information group (MIG) capabilities and processes.[33] Figure 2.1 shows the IE operations command center from the concept of employment.

---

[32] Burns, undated, pp. 3–4.

[33] See Appendix B for more detail on this concept of employment and the term *IE operations*.

Figure 2.1
IE Operations Command Center

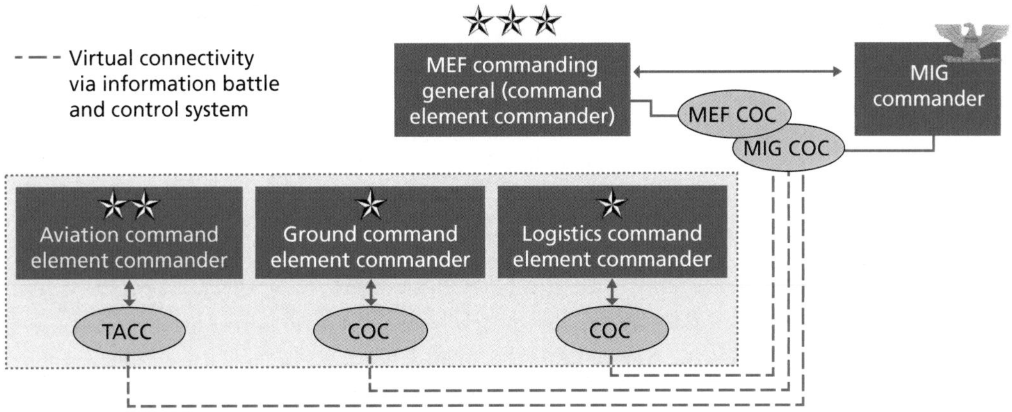

SOURCE: Adapted from U.S. Marine Corps, *Marine Air Ground Task Force Information Environment Operations Concept of Employment*, Quantico, Va.: July 6, 2017b, p. 7, Figure 4.

NOTE: COC = command operations center. TACC = tactical air command center. The MIG COC serves as the MIG commander's C2 center for IE operations and provides near-real-time IE operations battle management capabilities, as well as support to MEF planning efforts as directed. The MIG COC is physically or virtually connected to the MEF COC and any other MEF C2 center as required.

Games also provide an opportunity to integrate OIE with other tools of joint operations. The IE affects outcomes in the physical battlespace, whether or not combatants plan for their effects, and wargaming offers an environment where lessons may be learned without costing lives. Wargaming OIE can assist the Marine Corps in achieving a better holistic understanding of the IE, particularly by exploring the reactions of other actors and the underappreciated or unintended IE consequences of MAGTF operations.

There are also consequences to wargaming OIE poorly. Just as good wargames can expose important relationships in the IE and raise important questions in the minds of participants about causes, effects, and implications, poorly conducted wargames have the potential to impart negative learning and impress wrong lessons about the IE on participants. Games that imply that success in the IE comes merely from deploying the correct capabilities, that never force players to struggle with the blowback of nimble adversaries or unintended consequences of their own actions, or that exclude or fundamentally mischaracterize the IE all implicitly teach Marines that the IE is a peripheral consideration. Similarly, games that focus on combat as an exclusively physical undertaking, without considering the psychological impact of various battlefield events and actions, miss something important and risk teaching incorrect lessons.

The Current Status of OIE in Wargaming

Unfortunately, the existing literature and most common wargaming practices and tools do not support or substantially include OIE. Attrition-based defeat mechanisms predominate in the current generation of defense wargames, and most games do not adequately account for the IE.

For example, many joint and service games are based on or derived from approved support for strategic analysis (SSA) scenarios, concepts of operations, forces, and baselines. SSA products are developed collaboratively by the Office of Cost Assessment and Program Evaluation in the Office of the Secretary of Defense, the Office of the Under Secretary of Defense for Policy, and the Joint Staff, and they reflect combatant command plans and force management decisions.[34] The purpose is to give the services the opportunity to develop required capabilities and other products from a common set of departmental scenarios and assumptions. However, SSA products do not necessarily prompt consideration of IE dynamics, despite significant organizational pressure for DoD stakeholders to base their wargames on them.

The broader shift in attention to great-power competition with such potential adversaries as Russia and China after years of counterinsurgency can also prompt sponsors and participants to focus on conventional warfighting through wargames that continue to omit or downplay the considerable IE dynamics at play in even the most conventional conflicts.[35] There is a risk that this focus on kinetic, peer-oriented conflict will predominate despite the likely significant role that IE elements will play in conflicts with these near-peer adversaries (Russia, in particular) and increased attention to "gray-zone" competition that is ambiguous in nature.[36] There is room for improvement in the extent to which SSA products consider information and its operational impact, and these improvements could have important downstream effects as they are adopted by other DoD organizations. However, the tendency to ignore the IE and exclusively favor the physical aspects of warfighting permeates not just wargames related to SSA but most wargames (and the tools that support them).[37]

---

[34] Paul K. Davis, *Capabilities for Joint Analysis in the Department of Defense: Rethinking Support for Strategic Analysis*, Santa Monica, Calif.: RAND Corporation, RR-1469-OSD, 2016, p. 67.

[35] Then–Secretary of Defense James Mattis commented on the return of "great power competition" during the rollout of the 2018 national defense strategy. See James N. Mattis, Secretary of Defense, "Remarks by Secretary Mattis on the National Defense Strategy," transcript, U.S. Department of Defense, January 19, 2018.

[36] See U.S. Department of State, International Security Advisory Board, *Report on Gray Zone Conflict*, Washington, D.C., January 3, 2017, p. 1.

[37] Games focused on C2 issues have drawn greater attention to certain aspects of the IE. The U.S. Navy, in particular, has devoted considerable attention to C2 gaming—ranging from theoretical treatments in the early 2000s to concerted considerations of alternative C2 structures in its 2013–2016 global wargame series. Recent interest in multidomain C2 has brought another surge of attention to these topics, focusing on issues ranging from risk identification to concept development and the security of logistics lines of control.

One underlying reason for this failure to put doctrinal concepts into wargaming practice is that the literature on techniques to represent OIE in games is underdeveloped. This is partly the result of the same structural factors that cause gaming to be underdocumented. Wargaming is a practitioner's field, with no dedicated supporting academic discipline and no traditional hot houses for documented theory and recorded practice. Instead, most writing on games is focused on areas of game sponsor interest or analyzing the results of a specific game and their implications. Although there are reports on games related to OIE and specific IRCs, as we discuss in the next chapter, they often lack details on methodological approaches or an accounting of game data that is sufficient to support replication or independent evaluation of the quality of OIE representation. In addition, many of these reports are not publicly available—a problem that may be particularly acute for sensitive operational topics, such as cyber and electronic warfare. Given that there is not much applied wargaming literature of any kind, the dearth of writing on OIE in wargaming is not surprising.

The few texts that do exist tend to fall into a few broad categories. A small number of recent articles and historical critiques address the treatment of the IE and related issues. There is also a set of general texts on game design, targeting either the professional military gaming community or the broader commercial gaming sector, that discuss the role of information in games. Although these texts can be helpful, they do not specifically discuss the depiction of the IE and thus are likely too abstract for a designer who hopes to incorporate OIE into a game. Briefly, most of the discussion of the role of information in games focuses on various approaches to controlling and conveying situational awareness to players or on increasing verisimilitude in terms of C2 and access to communications. Overall, the literature allows that strong treatments of the IE should be important in games, but it does little to help steer a designer toward concrete practices.

When it comes to general gaming practices, practitioners regularly reference a core set of handbooks and guides to gaming fundamentals. These range from general texts on wargaming from both a professional and hobbyist perspective to handbooks

Although these games have not fully incorporated OIE, they consider, to some extent, the role of information, perceptions, and deception and thus are important sources of guidance on potential techniques and best practices for OIE gaming. Specifically, techniques to create barriers to player communication, including physical separation, time delays in communication, and degraded communication, are relevant to simulating certain aspects of OIE in games. See Robert C. Rubel, Director, Research and Analysis Division, Wargaming Department Center for Naval Warfare, U.S. Naval War College, "Using Wargames for Command and Control Experimentation," paper presented at the 8th International Command and Control Research and Technology Symposium, Washington, D.C., June 2003; Marc Selinger, "Air Force Eyes New Wargame on Multi-Domain Command and Control," *Defense Daily*, June 6, 2018; Brien Alkire, Sherrill Lingel, and Lawrence M. Hanser, *A Wargaming Method for Assessing Risk and Resilience of Military Command-and-Control Organizations*, Santa Monica, Calif.: RAND Corporation, TL-291-AF, 2018; and Heath and Svet, 2018.

from different wargaming centers.[38] Many of these texts focus on the management of information in games, stressing that designers have a great deal of freedom to control what information is available when and to which players as they simulate a particular scenario. However, these texts rarely make the jump to discussing the simulation of any of the aspects of the IE, much less the simulation of OIE. Put bluntly, designers are told that they have a powerful tool at their disposal but are given few guidelines as to its use.

The absence of formal literature should not be read as an absence of thinking and public discussion among wargamers on how to improve representations of the IE, OIE, and IRCs in games. For example, the Military Operations Research Society, a major hub of the wargaming community, ran a special session on cyberspace wargaming and analysis in October 2018.[39] Similarly, the 2018 Connections Wargaming Conference featured a "game lab," in which participants were able to propose their own topics of interest. Of the 30 topics proposed, at least seven related directly to an element of OIE.[40] We hope that this report will be one of many to draw from those types of conversations and improve the state of the literature.

Because the IE is important, because wargaming the IE is important, and because current common practice and guidance for wargames predominantly ignores the IE, the Marine Corps and the joint force need to seek improvements. To help build this foundation, the next chapter lays out requirements for the inclusion of the IE and OIE in wargames.

---

[38] The most notable of the first type is Peter Perla's *The Art of Wargaming*, originally published in 1990 (see Perla, 2011). Other frequently referenced texts include James F. Dunnigan, *Wargames Handbook: How to Play and Design Commercial and Professional Wargames*, 3rd ed., Lincoln, Neb.: iUnivers, 2000, and Francis J. McHugh, *U.S. Navy Fundamentals of War Gaming*, 3rd ed., reprint, New York: Skyhorse Publishing, 2013, originally published in 1966.

Examples of prominent texts from wargaming centers include Shawn Burns' *War Gamers' Handbook* (Burns, undated), the *U.S. Army War College's Strategic Wargaming Series Handbook* (Markley, 2015), and the UK Ministry of Defence *Wargaming Handbook* (UK Ministry of Defence, 2017).

[39] Military Operations Research Society, "Special Meetings," webpage, undated b.

[40] Author correspondence with the organizers, October 22, 2018.

CHAPTER THREE

Requirements for Wargaming Operations in the Information Environment

In this chapter, we discuss the requirements for OIE and the IE in wargaming—first in general and then as specific to the Marine Corps. Our intent in approaching the issue in this way is to think about wargaming requirements from a range of perspectives, moving from the general to the specific.

To that end, we discuss what we consider to be requirements for wargaming OIE. We found that many of these requirements are quite fundamental: There are significant requirements for basic theory, cause-and-effect, and other foundational knowledge or hypotheses that come before specific wargaming recommendations can be implemented.

General Requirements

Information as a term encompasses a wide range of subjects, and the role of information in warfare and other military operations is similarly broad. To identify shortfalls in how information and the IE are captured in wargames, we need to parse some of the disparate considerations that gather together under the broad tent of *information*. By understanding the ways in which information is relevant in warfare, we can bound general requirements for including relevant aspects of the IE in wargames.

Six Categories of Information

Based on experience, other research we have conducted, and interviews with stakeholders, we identified at least six key types of information that can be relevant to military operations and thus might need to be represented in wargames (depending on the game's objectives and level of abstraction):

1. situational awareness and situational understanding (including battle damage assessment)
2. C2, including communication
3. C2 warfare (C2W) and other factors that degrade situational awareness or C2

4. information or aspects of the IE that can cause subordinates to behave in ways contrary to the commander's orders or preferences
5. efforts to protect against contrary subordinate behavior or to encourage such behaviors among adversary forces (information for effect)
6. factors in and through the IE that affect the perceptions or behaviors of relevant actors other than adversaries.

Each category is summarized in Table 3.1 and described in greater detail in the sections that follow.[1]

Category 1: Situational Awareness and Situational Understanding

The first category of information in warfare is information about the operating environment or battlespace: where one's own forces are, where enemy forces are, where other relevant actors are, the state of those actors or forces, and what features of the environment might affect operations. This is commonly described as *situational awareness* or *situational understanding*, and, during actual operations, it is opposed by the natural forces of uncertainty collectively called *the fog of war*. Sensors and intelligence, surveillance, and reconnaissance (ISR) are capabilities that allow forces to fight through the fog of war. A subelement of this category includes monitoring changes in the state and

Table 3.1
Six Categories of Information

Category of Information	Description
1. Situational awareness and situational understanding	Information about one's own forces, adversary forces, other actors, and the environment that might affect operations
2. C2, including communication	Information that affects C2, including the availability and content of communications
3. C2W and other factors that degrade situational awareness or C2	Information warfare or attacks on situational awareness or C2 capabilities
4. Information that causes subordinates to behave in ways contrary to the commander's orders or preferences	Contradictory information, lack of communication, or behavioral factors leading one's own forces to disobey or act contrary to a commander's orders or preferences
5. Information for effect	Information that protects subordinate behavior or efforts to spread contradictory information, limit information, or otherwise encourage adversary forces to act contrary to their commander's orders or preferences
6. Factors that affect relevant actors other than adversaries	Factors in and through the IE that affect the perceptions and behaviors of relevant (non-adversary) actors (and efforts to influence other relevant actors)

[1] For examples of all six categories of information as part of Marine Corps OIE, see Christopher Paul and William Marcellino, *Dominating Duffer's Domain: Lessons for the U.S. Marine Corps Information Operations Practitioner*, Santa Monica, Calif.: RAND Corporation, RR-1166-1-OSD, 2017.

status of various forces (particularly enemy forces) that have been subjected to battlefield effects. The traditional version of this subelement is *battle damage assessment*, in which targets struck by various forms of fires are evaluated to determine the effectiveness of those fires and whether the targets need to be struck again. Note that this category includes information about the entire operating environment, not just the IE.

Category 2: Command and Control

The second category of information concerns C2 (including communications), especially the ability to communicate and transfer changes in understanding and instructions. This category recognizes the truth that a commander in a headquarters does not know about contact with an enemy formation at the exact moment the first scout in the field spots the first sign of movement; it accounts for delays that are often inherent in the system as new ISR data are processed, exploited, and disseminated, as well as delays that occur as orders are developed and communicated to subordinates. This category of information depends on networks and nodes, communication radii, chains of command, numbers of echelons, and so on.

The relevant content in this category can diverge at different echelons, in terms of both what is known where and what is relevant where. For example, a tactical formation will have ample information about its immediate environment before that knowledge is available at headquarters, and headquarters will know more about other areas of the battlespace sooner than personnel in a tactical formation. Similarly, exactly what is known and what is interesting will vary; at the tactical level, the precise location of enemy forces and details of terrain (such as whether terrain provides actual cover or just concealment) is of much greater interest, whereas at headquarters, broader brush strokes of data on the formations to which enemy troops belong and the location and disposition of a formation is of greater interest.

Category 3: Command-and-Control Warfare

The third category is a subset of what is sometimes called *information warfare* and represents an important transition in these categories of information—from describing what forces need to operate (the first two categories) to describing fighting with (or against) information. This category includes attacks on the functions in categories 1 and 2, as well as C2W and other attacks on situational awareness/situational understanding or the systems that convey that knowledge. In other words, it is about using OIE and IRCs to interfere with enemy C2 or to amplify the fog of war and either promote general uncertainty or lead enemies to incorrect conclusions about some aspect of the location, disposition, or possible courses of action of friendly forces.

Category 4: Information That Causes Subordinates to Behave in Contrary Ways

The fourth category concerns information or aspects of the IE or operating environment that can cause subordinates to behave in ways contrary to a commander's orders or preferences. This exposes another important relationship between information and the operating environment—namely, how information affects behavior. Why might subordinates not do what a commander wants? There are numerous possible reasons. Subordinates might not know what a commander wants because of failures in C2 (category 2) or because of inflicted failures in C2 (category 3). Subordinates in receipt of orders might be incapable of following them (lacking sufficient fuel or ammunition or having sustained so much damage as to be physically disrupted), but the commander may not know that because of failures in situational awareness. Whether correctly or incorrectly, subordinates might perceive a situation differently from the commander and thus act in accordance with the principles of mission command and in a way that is consistent with the overall commander's intent and the subordinates' perception of the situation. Subordinates might also act in contravention of the commander's wishes due to factors that are not strictly rational and are governed by psychology or emotion. This could be the baseline personality and proclivities of a subordinate (e.g., bold, timid, reckless) or due to effects wrought by such battlefield circumstances as distraction, suppression, panic, fear, shock, surprise, or rage.

Category 5: Information for Effect

The fifth category of information captures actions in or through the IE to mitigate or counter the effects in the previous category on one's own forces or to intentionally inflict such effects on adversary forces. We call this *information for effect*, and it encompasses a range of possible OIE, including efforts to harness the inherent informational aspects of military operations, as well as the employment of various IRCs, to affect and influence the behavior of enemies.

Category 6: Factors that Affect Relevant Actors Other Than Adversaries

The sixth and final category consists of factors in and through the IE that affect the perceptions or behaviors of relevant actors other than adversaries—basically, category 5 activities against targets other than enemy forces. Targets could include other actors in the immediate operating environment (such as nonstate actors, relevant civilian populations, or partner-nation forces), or relevant actors outside the area of physical operations (such as the domestic constituencies that support the adversary, one's own domestic constituents, senior leadership or national command authorities on either side, or citizens and leaders in nations not party to the conflict that contribute to a conflict's international legitimacy). This category is similar to categories 4 and 5, but its scope is broader, not only geographically but in terms of the types of relevant actors and timescale. Furthermore, whereas categories 4 and 5 were more focused on factors that affect actions and behaviors in combat, category 6 includes factors that affect perceptions and behaviors more broadly and over time. Thus, this category requires tracking and understanding baseline attitudes, legitimizing processes, and narratives drawn from the perspectives of those affected.

We highly recommend enlisting expert knowledge when designing games to capture this category of information, especially from experts in regional studies, conflict dynamics, social movements, narrative, public perception, public polling, or other relevant fields. These are specialized skill sets and generalists who are not familiar with these topics and methods for capturing them will not produce valid results.

The Six Categories of Information and Their Implications for Wargaming Requirements

In a wargaming context, adequately capturing these categories of information requires different mechanics. For example, when it comes to category 1 (situational awareness), a significant number of wargames dispense entirely with the fog of war, assuming perfect ISR and rendering this category of information (and several others) irrelevant. To remedy this tendency, a game requires mechanisms to track ground truth (where units actually are) and some mechanism to depict to players only the unit locations that they are supposed to know based on their own units' ISR capability and ability to detect others.

Representing category 1 information also requires a means to determine when to reveal to players the changing strength and status of various units. Category 2 (C2 and communications) requires a general mechanism for imposing delays on updates to orders and awareness, perhaps accompanied by some special mechanism to degrade communications or prevent units from receiving communications. However, on some timescales (perhaps a turn-based system in which turns represent hours), anything other than extreme delays could be abstracted away. Category 3 (C2W) requires additional complexity in the mechanisms used to capture categories 1 and 2 to allow actions that reduce detection ranges or probability. (Such actions could be anything from camouflage to the capability to jam or destroy sensors.) These mechanisms must allow players to attempt to execute relevant C2W activities as game actions, adjudicate the effectiveness of these moves, and have an impact on situation awareness (category 1) and C2 (category 2) in the game. Furthermore, games involving efforts at concealment and deception ideally should have a recording mechanism to track positions and play so that players who fall victim to well-executed concealments or deceptions can go back in time and see what happened, increasing the chances that they will accept the verisimilitude of the game rather than perceiving it as unfair (and thus unvalued as a learning experience).

Mechanisms that capture category 4 (factors that cause subordinates to behave in ways contrary to the commander's orders or preferences) must respect the possibility that subordinates will do something other than what is ordered. Such a mechanism could follow a simple sequence: When a unit is ordered to act or when it is otherwise time for a unit to act, the mechanisms for categories 2 and 3 could be consulted to see whether the unit has received the orders (in whole or in part). If not, a mechanism for actions in the absence of orders could be consulted, which might consider last orders received, commander's intent, and the unit leader's proclivities and perception of the situation. If orders *were* received, then a slightly different mechanism could check for factors that might prevent following orders, such as incapacitation, suppression, or more current situation awareness that contradicts the orders received. Absent any such factors, orders would be followed. However, if such factors are *present*, the game could employ a probabilistic test against the strength of those factors to determine whether orders were followed.

Category 5 (countering category 4 effects or intentionally inflicting those effects on adversaries) is mostly covered by the mechanisms necessary for category 4, with some modest additions. Category 5 also requires a mechanism for adjudicating the effects of efforts to change conditions faced by a unit either favorably or unfavorably, which would then circle back to the mechanisms used to adjudicate category 4.

Category 6 (factors that affect relevant actors other than adversaries) requires a fairly significant set of additional mechanisms. In the first five categories, the scope of effects was constrained primarily to the impact of information and the IE on military units and the players controlling them, as well as combat operations and geographic battlefields or battlespaces. Although contextual factors weigh into how a wargame accounts for the first five categories of information, they are not the primary focus. Here, however, the opposite is true. Capturing category 6 requires a mechanism to track various relevant actors, their conditions, the impact of other game actions on those conditions (including actions to influence them and other sources of influence, such as collateral damage), the possible range of actions these actors might take, and how those actions might affect the game.

Conceivably, game designers might choose to partially obscure some of these category 6 mechanics so that players know, for example, that disgruntled noncombatants might riot or protest but not how likely that is, how severe riots might become, what second-order effects might result, or exactly how each game action might affect those outcomes. In terms of how relevant actors' actions might affect the game, the range of possibilities is quite broad and would need to be constrained for a game to remain manageable. That said, such actions could plausibly have a very significant impact on a game. One could imagine a game design in which one side's national will to fight, governed by domestic public opinion, had a built-in random constraint governing how long the game could last—with declining public support resulting in the game ending (troops are withdrawn), even though that side's forces were tantalizing close to achieving mission objectives.

Marine Corps Requirements

Wargaming OIE and the IE requires, at least notionally, the representation of at least some elements of the six categories of information. In this section, we move from general (and generally abstract) requirements to those that are specific to the Marine

Corps, beginning with requirements derived from our interviews with stakeholders and moving to concrete considerations specific to Marine Corps OIE-related capabilities.

Stakeholder Requirements

We held discussions about requirements with the research sponsor, individuals who operate within specific IRCs, and members of the wargaming community who were familiar with Marine Corps wargaming, training, or exercises related to the IE. Although these interviews captured a limited fraction of the full range of possible Marine Corps perspectives on the topic, those with whom we spoke were known by the sponsor or by RAND to have focused expertise and relevant stakeholder knowledge. These interviews brought to light a range of requirements, from deeper IE scenarios and improved adjudication of IE effects to a full consideration of the cognitive realm. Some appeared simple to address, but others would entail significant effort and investment.

Beyond the simple need to include IE considerations in wargames, stakeholders identified a need for depth in representing the IE-related scenarios that Marines encounter. One stakeholder explained that the Marine Corps has been in a "transition period" since the MIGs have been stood up, noting that there has been limited time to develop IE-rich scenarios. However, that same stakeholder also identified a need to "force people out of their comfort zones" through such scenarios and to confront situations when attempts to achieve effects in and through the IE do not work. For example, if SIGMAN or C2 did not work as planned, what should happen next in the game?[2] Related to deeper IE scenarios was a desire for better representation of national assets and coalition capabilities that could also affect the IE, as well as the representation of other important actors, such as coalition partners, international organizations, and nongovernmental organizations.

Another requirement concerned appropriate feedback on IE-related actions during a wargame. Stakeholders cited such challenges as not receiving a response to requests for information during games regarding actions in or through the IE and a need for adjudication of IE effects that was more rigorous than "ballparking" a response within the game.[3] Adjudication quality reportedly varied: Sometimes, well-versed experts were enlisted to adjudicate IE effects; in other instances, IE adjudication was (poorly) handled by "pick-up" teams without relevant experience or formal military occupational specialty training.[4] One stakeholder saw a need to move away from the unstructured discussion that traditionally characterized some Marine Corps wargaming events to

---

[2] Phone interview with a Marine field-grade officer involved with MIG exercises, March 21, 2018.

[3] Phone interview with a Marine field-grade officer involved with MIG exercises, March 21, 2018; phone interview with a Marine field-grade officer involved in information warfare, March 30, 2018.

[4] Phone interview with a Marine field-grade officer involved with MIG exercises, March 21, 2018.

more-structured, even rigid, adjudication.[5] While the need for *some* kind of IE feedback in a game may be relatively easy to meet, well-researched and well-justified IE adjudication requires a more systematic understanding of the IE and human dynamics, as well as an investment in developing a bench of adjudicators with deep expertise. Because poor adjudication can result, at worst, in false confidence and negative learning, knowledgeable IE adjudication is a core requirement for improving representation of OIE and the IE in wargames.

Related to this idea of understanding the effect of various actions on the IE was the question of breakpoints: the casualty level or other circumstances at which a unit or formation ceases to fight and "breaks," either fleeing, cowering, or surrendering. One stakeholder raised the question of how IRCs affected or did not affect the breakpoint in a wargame, and how they might do so.[6] This is a significant consideration, and we explore breakpoints, defeat mechanisms, and will to fight in wargames in more detail in Chapters Five and Six.

Another identified requirement was meaningful inclusion of the cognitive aspects of the IE, particularly decisionmaking. One stakeholder noted that simple combat modeling would be sufficient if the IE had only physical and informational characteristics, but because the IE includes a cognitive dimension, simple models are inadequate.[7] The requirement to explore the *cognitive* aspects of an operation points to another requirement for wargames that treat the IE as a substantial element of the battlefield: There is need to structure games so that players have the freedom to substantively change their decisions based on IE developments and their perceptions of the information they send or receive. Players need to be given the latitude to make significant decisions, but they should also be required to live with the consequences of their actions over the course of the game. Seminar-style games and events structured like planning exercises are insufficient to meet this important requirement.[8]

Instead, there is a need for better understanding of player mental models, beliefs, perceptions, interpretations of information presented in a game, and the effects of information on decisionmaking.[9] These considerations constitute the cognitive underpinnings of a wargame, one that is shaped by the players themselves (either what they bring to the game or what they perceive and experience during game play). These constructs, created by the players, mediate their understanding of what in going on in the world of the game; game designers, sponsors, or even other inputs into the game

---

5 Phone interview with a Marine field-grade officer involved in information warfare, March 30, 2018.

6 Phone interview with a Marine field-grade officer involved in information warfare, March 30, 2018.

7 Phone interview with a Marine field-grade officer involved in information warfare, March 30, 2018.

8 See Appendix A for a description of seminar-style wargames and a summary of differences between this approach and other types of wargames.

9 Unfortunately, in most games, this is handled in an informal or ad hoc matter at best. However, it becomes critical when information is a key component of decisions and effects in the wargame.

exert a limited amount of influence on these cognitive frames. However, the cognitive constructs within a game serve as imperfect conceptual models that may or may not be generalizable to the real world; adversaries and other actors may not perceive or make decisions in a game in the same way that U.S. players would. There is therefore a requirement for better qualitative analysis of these constructs and mental models, as well as a comparison between how they are represented in games and how they are likely to manifest in the real world.

Table 3.2 summarizes our discussion of stakeholder-identified requirements, the implications for wargaming, and the underlying support necessary to adequately address those implications.

An enormous amount of knowledge and understanding is required to adequately support the wargaming implications in Table 3.2, including conceptual models or hypotheses for cause and effect in the IE that are explicitly articulated and supported by evidence drawn from data on causal relationships, realistic feedback within the wargame, and analytic methods that improve understanding of players' mental models and decisionmaking in the game.

Information-Related Capabilities as a Source of Requirements for Wargaming OIE

In addition to the requirements specified or implied by Marine Corps stakeholders, there is another obvious source of requirements for wargaming OIE: the various capabilities that the Marine Corps can employ to conduct OIE. Although the *Joint Concept for Operating in the Information Environment* calls out the importance of the inherent informational aspects of all military activities, certain capabilities are explicitly and primarily intended to generate effects in and through the IE. These are traditionally referred to as *IRCs*. We note "traditionally" because the latest revision of JP 3-0, *Joint Operations*, instead discusses "joint force capabilities, operations, and activities for leveraging information."[10] Under that heading, JP 3-0 lists the follow-

Table 3.2
Stakeholder-Identified Requirements for Wargaming the IE

Stakeholder-Identified Requirement	Wargaming Implications
Depth in IE scenarios	• More deeply developed IE backstory • Additional actors in the IE represented in the game
Appropriate IE feedback	• Wargame mechanisms and improved adjudication of IE effects
Representation of the cognitive aspects of the IE	• Wargaming formats with adjudication and multiple moves • Qualitative analysis of player mental models and constructs • Comparison of games' cognitive constructs with those in the real world

[10] JP 3-0, 2018, p. III-22.

ing functions: key leader engagement, public affairs, civil-military operations, military deception (MILDEC), military information support operations (MISO), operations security (OPSEC), electronic warfare, combat camera, space operations, special technical operations, cyber operations, DoD information network operations, cyber-enabled activities, and commander's communication synchronization.

Marine Corps doctrine still uses *IRCs* but intentionally does not provide a list; the definition of the term *IRC* is open-ended, such that capabilities outside those that could be listed might be IRCs under certain circumstances. An IRC is "a tool, technique, or activity employed within a dimension of the information environment that can be used to create effects and operationally desirable conditions."[11] Although there is no doctrinal list, informal lists abound within the Marine Corps. One stakeholder described charts listing as many as 30 distinct IRCs. Another provided us with a list that included the following "information environment activities" for the Marine Corps: intelligence, C2, cyber operations, MISO, civil-military operations, space operations, electromagnetic spectrum operations, communication strategy and operations, OPSEC, information assurance, physical security, key leader engagement, defense support to public diplomacy, physical attack, and MILDEC.[12] That same respondent added SIGMAN as a notable omission to the list, probably because it is a nascent concept and capability. Because the Marine Corps has all of these capabilities, there is an implicit requirement to be able to include and meaningfully represent them in wargames.

To connect the general requirements discussed so far with Marine Corps–specific requirements, we notionally mapped the six categories of information to these IE activities or IRCs. Table 3.3 presents the results of this mapping exercise. Any wargame designed with a goal to meaningfully include all IRCs will need to meet the requirements of all six general categories of information. Games that include only a subset of IRCs may need to cover only a subset of the categories, safely ignoring or abstracting the remainder.

The general requirements for including the IE in wargaming, the specific requirements drawn from discussions with stakeholders, and a review of the various capabilities that the Marine Corps considers explicitly tied to OIE led us to develop the following summary of requirements:

- game mechanics that accommodate all six categories of information
- games that include IE-related scenarios
- games that represent OIE in greater depth or with more realism
- mechanisms that incorporate effects from OIE on noncombatant populations

[11] JP 3-13, *Information Operations*, Washington D.C.: U.S. Joint Chiefs of Staff, incorporating change 1, November 20, 2014, p. I-3.

[12] Information provided in an interview with a Marine Corps civilian, April 16, 2018.

Table 3.3
Information-Related Capabilities and the Six Categories of Information

	Category of Information					
	1	2	3	4	5	6
MAGTF IE Operational Function	**Situational Awareness and Situational Understanding**	**C2 and Comms**	**C2W**	**Contrary IE Effects on Subordinates**	**Information for Effect**	**Factors That Affect Actors Other Than Adversaries**
Intelligence	X		X			
C2		X	X			
Cyber operations			X	X	X	
MISO			X	X	X	X
Civil-military operations						X
Space operations	X	X	X			
Electromagnetic spectrum operations			X	X	X	
Communication strategy and operations						X
OPSEC	X		X		X	
Information assurance	X	X			X	
Physical security	X	X	X	X	X	
Key leader engagement						X
Defense support to public diplomacy						X
Physical attack	X	X	X	X	X	
MILDEC	X		X	X	X	
SIGMAN	X		X		X	

- mechanisms that show the effects of noncombatant populations on the IE and the operating environment
- mechanisms that show the effects of other actors (e.g., coalition forces, partner-nation forces) on the IE
- games that provide players with feedback related to the IE and OIE
- ability to include assessment of OIE in games
- ability to provide in-game situational awareness of the IE (rather than just situational awareness of the spatial domains)
- games that capture the effects of the IE on combat
- game that include SIGMAN capabilities
- games that include deception or surprise (as well as OPSEC and SIGMAN)
- games that include cyber and other technical capabilities
- games that include electronic warfare
- games that include various forms of influence targeting adversaries or other relevant actors.

In Chapter Two, we argued that assumptions about the nature of conflict drive the identification of defeat mechanisms, which then drive wargame design. Wargame design, in turn, determines the adoption or development of wargame tools. The general lack of OIE and IE considerations in defense wargaming practice can be traced back to assumptions about the nature of conflict that do not adequately factor in information considerations: Specifically, most wargames assume attrition-based defeat mechanisms. Given those assumptions, it is no surprise that few existing wargaming tools are appropriate for representing OIE and the IE. These upstream issues are more important than whether wargaming tools can perform these functions. What people assume about the role of information and its implications in warfare, and the resulting defeat mechanisms that they build into games, will ultimately drive how wargaming tools are used and whether they will be used in ways that appropriately represent the importance of the IE—and whether wargames will meet the requirements identified in this chapter.

Building on this discussion, Chapter Four evaluates several assumptions about the nature of conflict that properly emphasize OIE. Chapter Five identifies a series of defeat mechanisms beyond simple destruction or attrition. These two chapters attempt to fill in the gap common in wargame design and offer useful ideas for wargame design going forward.

CHAPTER FOUR

Assumptions About the Nature of Conflict

In this chapter, we consider assumptions about the nature of conflict and examine the associated role of OIE. As discussed at the end of Chapter Two, current wargaming approaches focus on the kinetic aspects of war. To develop the foundations of a more balanced approach, we examined wargaming theory that addresses spheres of war beyond the physical, in which OIE and the IE are critical components of warfare. We paid particular attention to the implications for military objectives, including battlefield objectives. Given Marine Corps equities, this discussion focuses on ground combat, but it should be noted that many of these same phenomena are relevant to conflict in other physical domains.

As part of our look at military theory, we revisit the mental, morale, and physical dimensions of war encapsulated in many important military writings. Understanding war through these three spheres shows us how thoroughly woven information considerations are into the major aspects of warfare. We argue that the IE is relevant to all three spheres but that it has the greatest impact on military operations through the mental and morale spheres (as those exist predominantly within the IE).

We begin the discussion with first principles, considering objectives in warfare, the nature of defeat in combat, and the role played by mental and morale factors (and thus the IE) in the outcomes of military operations and battles. If the IE is important in war, then it should feature prominently in wargames.

Military Theory

Objective in Warfare

What are the objectives of warfare and combat?[1] Carl von Clausewitz tells us that war is politics by other means, that war employs force to compel the enemy to do our will,

[1] We intentionally constrained the scope of the objectives to warfare objectives. Certainly, there are IE-related objectives in operations other than war, including stability operations and humanitarian assistance and disaster relief. And there are IE-related objectives in various forms of low-intensity conflict, such as counterinsurgency and counterterrorism. Games that pursue informational objectives with information capabilities constitute one

and that to accomplish this objective we must render the enemy powerless.[2] Some have understood that to mean that the objective of war is to destroy the enemy. Others have criticized destruction as missing the mark, instead noting, in the words of one theorist, "The art of war consists largely of learning how to defeat the enemy *without* complete destruction."[3] Still others have noted that destruction is not an end unto itself but that it is a useful tool, because it can create advantages or other opportunities.[4]

Considered more systematically, our review of military theory revealed three major strands of military objectives. The first is the higher-level objectives, the observation akin to Clausewitz's "politics by other means," that war seeks to impose or enforce a policy of some kind.[5] The second is that to succeed in imposing policy, it is often prerequisite (though not always) that the friendly force *defeat* the enemy.[6] We emphasize *defeat* here because we found a wide range of interpretations of what it might mean to defeat an enemy and how one might go about doing so, and the range and scope of those variations is quite informative. The third major strand of objectives is to gain an advantage over the foe. Achieving a position of advantage is military wisdom supported across the spectrum and history of military theory and is the foundation of the maneuverist approach to warfighting. Advantage can come from a wide range of situations or activities. According to Robert Leonhard,

> The physical maneuver of troops on the battlefield is just one method among many to achieve the advantage. The *goal* of obtaining an advantage over the enemy can be served *by* various *means*. The means will change from war to war, or even from day to day, but the underlying concept of *advantage* remains. It is this immutable idea of advantage that should be codified as a principle, not the transitory ways to get the advantage.[7]

"Advantage" is nearly ubiquitous in military thinking and is clearly a good thing to have. However, we are skeptical of its utility as an objective. Rather, we prefer to think of advantage as a means to an end rather than an end unto itself. In much the

---

class of problem, but, here, we focus on how information contributes to warfighting objectives to support the design of games that realistically simulate the pursuit of warfighting objectives using both physical and informational power.

[2] Carl von Clausewitz, *On War*, Michael Howard and Peter Paret, eds., Princeton, N.J.: Princeton University Press, 1976.

[3] Robert R. Leonhard, *The Principles of War for the Information Age*, New York: Ballentine Books, 1998, p. 77.

[4] Jim Storr, *The Human Face of War*, London: Continuum, 2009.

[5] See Leonhard, 1998, and Emile Simpson, *War from the Ground Up: Twenty-First-Century Combat as Politics*, Oxford, UK: Oxford University Press, 2012.

[6] Even Fuller's writings acknowledged that, sometimes, "[a] military victory is not in itself equivalent to success in war" (John Frederick Charles Fuller, *The Foundations of the Science of War*, London: Hutchinson, 1926, p. 76).

[7] Leonhard, 1998, p. 59.

same way that defeat of an enemy is a means that may or may not allow the victor to impose policy preferences, advantage is even more deeply subordinate. Having an advantage allows one to impose costs on the enemy at lower relative costs to one's own forces. Such costs may contribute to defeat or to an ability to impose those policy preferences, but seeking and finding advantage alone does not complete the equation of warfare. Wars, and wargames, do not end when one side has an advantage but, rather, when one side capitalizes on that advantage to defeat the other.

What Does it Mean to "Defeat" a Foe?

Given the centrality of *defeat* to war, and thus to wargaming, it makes sense to unpack the term before attempting to determine how OIE contribute to its accomplishment. We found numerous perspectives on defeat in military theory. One possible understanding is that *defeat* denotes the complete destruction of the foe's fighting forces. Although some level of destruction is implicit in many interpretations of defeat, few theorists explicitly call for complete destruction. This is because the preponderance of defeated military forces throughout history have been defeated long before every combatant has fallen as a casualty. While complete destruction of a force certainly would result in defeat, *defeat* still has meaning at lesser levels of destruction.

Of course, Clausewitz offers that it is possible to defeat an enemy by capturing or destroying its center of gravity. If that center of gravity is the enemy's army, then we fall back to a destruction argument. But if it is something else (a key city, a ruler), there is a way to defeat the enemy without much destruction. In the words of Emile Simpson, "The centre of gravity was simply what mattered to an opponent, or more specifically what could be made to matter."[8]

Many explorations of defeat relate to an opponent's will—specifically, the will to fight. Leonhard characterizes defeat as "a condition in which an enemy force has given up the desire to fight."[9] Jim Storr describes the most common resolution of combat as collective withdrawal from participation in combat.[10] He further notes the cascading nature of the withdrawal of participation: First, a few individuals cease to participate, eventually leaving their fellows in their squad or platoon little choice in the matter; then, the same cascade affects the next higher echelon, and then the next. As more and more of a commander's formations withdraw their participation, the commander is left in command of fewer and fewer troops and has fewer and fewer options. B. A. Friedman describes this reduction of will to fight as "breaking the moral cohesion of the opposing force," which ends that force's ability to function as an effective unit.[11] As to when that loss of will to fight, or withdrawal from participation in combat, or breach

---

8 E. Simpson, 2012, p. 132.

9 Leonhard, 1998, p. 211.

10 Storr, 2009.

11 B. A. Friedman, *On Tactics: A Theory of Victory in Battle*, Annapolis, Md.: Naval Institute Press, 2017, p. 17.

of moral cohesion, or defeat occurs, Storr notes that it "comes when, and only when, the enemy believes himself to be beaten."[12] Past work has identified a variety of factors that can influence unit-level will to fight, including training, support, and cohesion, with each having a different influence, depending on the context.[13]

Defeat by breaking the will to fight can apply not only to the troops and formations engaged in fighting but also at the national level.[14] If the head of state decides that the cost of war can no longer be borne, or if antiwar protests or casualty levels undermine public confidence, forces that are otherwise still willing to fight can be withdrawn and, thus, in some sense, defeated. As with unit-level will to fight, scholars have emphasized that these factors will have different valence and magnitude of impact, depending on the specific conditions of a conflict.

The descriptions of defeat that we find most compelling described it as something that can be achieved through the combination of reducing an adversary's physical capacity and will to fight, recognizing that a reduction in one often corresponds to a reduction in the other.[15] In fact, even more compelling are the descriptions of the nature of war and defeat that follow the logic of J. F. C. Fuller and acknowledge three spheres of force: mental, morale, and physical.[16]

J. F. C. Fuller and the Mental, Morale, and Physical Spheres of War

While the notion of the centrality of three spheres (mental, morale, and physical) stems from Fuller's writings in the interwar period, we have seen them echoed and engaged by numerous other theorists, including John Boyd, Robert Leonhard, and B. A. Friedman, and in the capstone Marine Corps Doctrinal Publication 1, *Warfighting*.[17]

---

12 Storr, 2009, p. 3.

13 Ben Connable, Michael J. McNerney, William Marcellino, Aaron Frank, Henry Hargrove, Marek N. Posard, S. Rebecca Zimmerman, Natasha Lander, Jasen J. Castillo, and James Sladden, *Will to Fight: Analyzing, Modeling, and Simulating the Will to Fight of Military Units*, Santa Monica, Calif.: RAND Corporation, RR-2341-A, 2018.

14 Michael J. McNerney, Ben Connable, S. Rebecca Zimmerman, Natasha Lander, Marek N. Posard, Jasen J. Castillo, Dan Madden, Ilana Blum, Aaron Frank, Benjamin J. Fernandes, In Hyo Seol, Christopher Paul, and Andrew Parasiliti, *National Will to Fight: Why Some States Keep Fighting and Others Don't*, Santa Monica, Calif.: RAND Corporation, RR-2477-A, 2018.

15 Antulio J. Echevarria II, *Military Strategy: A Very Short Introduction*, Oxford, UK: Oxford University Press, 2017.

16 Fuller, 1926.

17 John Boyd, *A Discourse on Winning and Losing*, Grant T. Hammond, ed., Maxwell Air Force Base, Ala.: Air University Press, 2018; John Boyd, "Patterns of Conflict," briefing slides, December 1986, Slide 137 (also see John Boyd, "Patterns of Conflict," video series with accompanying reconstructed slides posted by Dan Grazier, Project on Government Oversight, 2015); Leonhard, 1998; Friedman, 2017; Marine Corps Doctrinal Publication 1, *Warfighting*, Washington, D.C., June 20, 1997.

The physical sphere is the easiest to understand, being the realm of bodies of fighting troops, their equipment and other materiel, capabilities, capacity, ammunition, fuel, fortifications, and so on, along with their disposition and destruction. Some things that have physical components may have a greater impact in one of the other spheres. For example, fortifications may have a greater impact in the mental or morale spheres (by appearing daunting or unassailable) than through the actual physical protection they provide. Similarly, although position is a physical characteristic, it may bring advantages beyond those in the physical sphere (an unexpected position conveying surprise is one example). What constitutes physical damage remains clear and manifests as death, destruction, physical weakening, wounds, depletion, and other physically identifiable changes. And, of course, physical damage has secondary effects in the other spheres.

Physical damage is difficult to recover from. Some wounds heal but many require lengthy recovery times or are so substantial as to prevent troops from ever returning to combat duty. Some damaged equipment or vehicles can be repaired, but that too can take time and require supplies not readily on hand. However, personnel, materiel, and munitions can all be replaced. This process is called *reconstitution* and is generally considered part of the logistics process in which replacement personnel and materiel are brought forward and integrated. Units at rest can have their supplies replenished and receive replacements. The more damaged and depleted a unit, the longer it will take to restore to full strength, a result of both the additional time required to bring new personnel and materiel forward and the time required to assign and integrate replacements within the unit's hierarchy.

Fuller and other theorists write about the "moral sphere." As they describe it, this has relatively little to do with morality, per se, and is much more about *morale*.[18] To avoid confusion about the connotation, we refer to the *morale sphere* throughout this report (except when directly quoting from these works).

Fuller equates the morale sphere with determination; he specifically calls out *surprise* as an example of exertion of maximum morale pressure and *endurance* as the power to resist morale pressure.[19] For Fuller, the morale sphere is the domain of fear and courage. Friedman described the "moral aspects" of battle as including the "morale of the troops engaged, the cohesion of the military units involved, and their spirit and enthusiasm for the task."[20] If decisions and intentions inhabit the mental sphere, then the will to carry out those decisions belongs to the morale sphere.

---

18 Note that the term *morale* can include a range of meanings. Here, we are less concerned with the notions of "happiness" or "satisfaction" and focus more on morale as an indicator of cohesion, fighting spirit, courage, or endurance. See the discussion in Connable et al., 2018, pp. 43–47.

19 Fuller, 1926, p. 299.

20 Friedman, 2017, p. 21.

In addition to will, cohesion, fighting spirit, and morale, Friedman does note the role of morality in the morale sphere. He writes,

> The soldier or Marine who views his cause as just and ethical will have higher morale, and this will thus affect his enthusiasm, his discipline, and his tactical decisions. Units engaged in the pursuit of moral aims have higher moral cohesion.[21]

He further notes that ethical behavior in combat can affect this sphere, because violating such norms as maintaining proportionality and avoiding harm to noncombatants can cause troops to feel shame and guilt, reducing morale and diminishing moral power and cohesion.

Damage in the morale sphere includes weakening resolve, fear, and deteriorating morale and will to fight, all of which accumulate or have a cascading effect, potentially to the point that moral cohesion snaps and individual—followed by collective—willingness to participate in combat ends. Recovery from morale damage is not as straightforward as physical damage. Certainly, morale damage recovers over time, but to varying degrees and over varying amounts of time. Minor morale damage, such as fear, the experience of surprise, or discouragement after being suppressed by fires, generally disappears fairly quickly once the source of the damage is withdrawn (or troops withdraw from it). More significant morale damage may take longer to mend. For example, troops who lose moral cohesion and break may flee for some time. When they cease flight, they may rally and re-form into useful units, but their morale may remain fragile, and they may be prone to fleeing again if pressed. Some fraction of troops who lose moral cohesion and break may not recover within the span of a battle and may remain unfit for duty long after. Some kinds of morale damage can be more enduring: troops whose unit or formation has been previously defeated may carry depressed morale and remain hesitant for some time, perhaps until they have had a significant period of rest or battlefield success, or until many of the troops who experienced the defeat have rotated out of the unit. Similarly, troops who have been involved in unethical behavior or atrocities may carry that shame indefinitely.

The mental sphere is the sphere of imagination, judgment, planning, decision, intention, and direction. Fuller frames actions in the mental sphere as attacking the plans of the enemy commander.[22] Friedman describes action in the mental sphere as preying on the mind of the enemy commander through deception (to prevent accurate perception), surprise (to limit decisionmaking time), confusion (to corrupt the decision-making process), or shock (precluding any decision but fleeing).[23] Our impression is that the mental sphere covers both C2 and ISR—that is, the various inputs and

---

[21] Friedman, 2017, p. 93.

[22] Fuller, 1926.

[23] Friedman, 2017.

sources of input that shape perception, the resultant judgments and decisions, and how decisions are conveyed (usually as orders) to subordinates.

Although the mental sphere notionally covers the judgment and intention of all enemy combatants, Fuller particularly emphasized the enemy commander as the one whose mental deliberations are most consequential in war. He further emphasized the "mental conditions" that potentially affect the enemy commander, including habits, proclivities, adherence to doctrine, and education—anything that might provide insight into likely plans at any given time or indicate how the commander might respond to psychological stressors.[24]

Damage in the mental sphere could take a number of different forms. It might include decreasing ISR capability or decreasing awareness, or perhaps increasing misperception through intentional deception. It could include degradations in C2, either through casualty-induced breaks in the chain of command, cyber or electronic warfare effects that degrade lines of communication, or simply being out of communication with relevant commanders. It could also include the loss of the commander or even just the loss of new direction from a commander who is out of communication, distracted, or forced to focus on other matters (such as fleeing from an assault on the headquarters).

Damage in the mental sphere (as understood here) is usually remedied quickly when the conditions that caused the damage are remedied. When jamming or communication interruptions end, so too does the associated damage. When situational awareness is restored, so is decisionmaking ability. When a lost commander is replaced and the chain of command is restored, orders resume their flow and damage in the mental sphere rapidly diminishes. Of course, some damage in the mental sphere can linger longer. If a network attack corrupts computer systems, they may be down for some time. A commander and staff who have been deceived once may be more likely to be deceived again or may be more hesitant, fearing that they will be deceived again. A replacement commander may not be as familiar with subordinates, as confident, or as good at decisionmaking, resulting in some enduring damage in the mental sphere. (Then again, the removal of an incompetent leader could push a formation's effectiveness in the mental sphere to new heights.)

Fuller argued that the three spheres were an inseparable trinity, claiming, "Mental force does not win a war; moral force does not win a war; physical force does not win a war; but what *does* win a war is the highest combination of these three forces acting as one force."[25] Defeat, then, under this conception, occurs when a force is sufficiently depleted in any of the three spheres with what constitutes "sufficiency," determined, in part, by the characteristics of the force and, in part, by damage accrued in the other

---

[24] Fuller, 1926, p. 183.

[25] Fuller, 1926, p. 146.

spheres.[26] So, a force with perfect moral cohesion may well fight to the last casualty, and defeat may require complete physical destruction; however, it is very unlikely that a force could maintain perfect moral cohesion in the face of massive physical casualties, and, at some point, combined losses would be significant enough for the force to break. Similarly, if we consider two different forces, we would expect the one with greater initial moral cohesion to accept more physical damage before breaking, all other things being equal. Furthermore, a force that has sustained no physical damage at all might break if it loses communication with command (depleted mental sphere) or observes a continuous stream of panicked and fleeing friendly forces (a severe threat in the morale sphere).

This description of the three spheres of warfare gives us much clearer leverage to describe what is missing in many analytical depictions of war. By focusing almost exclusively on the physical sphere of kinetic conflict, the mental and morale are given short shrift. However, these two spheres are precisely where the majority of OIE takes place. To illustrate this point, we refer back to the six categories of information in OIE defined in Chapter Three:

1. Situational awareness and situational understanding (including battle damage assessment)
2. C2, including communication
3. C2W and other factors that degrade situational awareness or C2
4. Information or aspects of the IE that can cause subordinates to behave in ways contrary to the commander's orders or preferences
5. Efforts to protect against contrary subordinate behavior or to encourage such behaviors among adversary forces (information for effect)
6. Factors in and through the IE that affect the perceptions or behaviors of relevant actors other than adversaries.

Although physical actions (e.g., cutting cables) can be used to weaponize all these information types, the primary desired effects are in the mental and morale spheres (e.g., playing on the adversaries fear of being outflanked). For example, operations to shape situational awareness aim to change the adversary's decisionmaking processes. Similarly, C2 and C2W aim to shape the information available to commanders to make decisions. Similarly, efforts to shape the behavior of subordinates, prevent the adversary from destroying the morale of one's own forces, or shape the perceptions and behaviors of other populations all require actions in the mental and morale spheres. In short, if we are to represent OIE in a meaningful way, we must consider actions and effects in the mental and morale spheres.

[26] This argument may not hold for the extremes of terrorism and nuclear war; Fuller's focus is more conventional.

The Three Spheres in Wargames

If warfare is best described as a trinity of the physical, mental, and morale spheres, the question then becomes how all three spheres can be depicted in games. Wargames consider the three spheres to a greater or lesser extent, but tend to make the physical sphere paramount, when, arguably, morale is most important. Most critical for the purpose of this report, even when wargames include mental and morale considerations, they usually appear only as secondary effects of physical applications of combat power: A unit is suppressed or pinned by physical fires and may break or rout, but only as a consequence of casualties.

For wargames to meaningfully capture the impact of OIE, they must consider all three spheres and allow game actions other than physical movement and physical combat to have other-than-physical effects. Fortunately, the same sort of three-sphere adjudication mechanics that would allow OIE to be meaningfully represented in wargames will also capture the impact of physical combat actions even better than current attrition-based models by allowing destruction to have an explicit rather than implicit impact on the mental and morale spheres.

It is not a new observation that, to fully capture warfare, games must include more than kinetic effects. For example, in a review of Navy gaming written in 2004, the authors state,

> If we accept the notion of the three domains of real war—physical, informational, and cognitive—then the wargame designer must somehow condense that real universe into the game universe. He does this by combining the six dimensions of wargaming—time, space, forces, effects, information, and command—to form three interconnected topologies—operational, informational, and command. These topologies are the interfaces and engine through which the players enter and transform the universe of the game. The measure of the game's realism is how well the relationships the players have with the game topologies reflect the relationships real-world commanders have with the real domains.[27]

However, a review of gaming practice shows considerable shortfalls in the representation of the mental and morale aspects of warfare. Although there are many reasons for this lacuna, it is a gap in practice that must be bridged to achieve useful depictions of OIE.

If we use Fuller's construct of the mental, morale, and physical domains of war to categorize current games, most consider the physical sphere, almost to the exclusion of the mental and morale spheres. For example, a recent RAND survey of 62 military and commercial games and simulations found few military products that included "will to

---

[27] Peter P. Perla, Mike C. Markowitz, Christopher A. Weuve, Stephen Downes-Martin, Michael Martin, and Paul V. Vebber, *Transforming Naval Wargaming: A Framework for Operational-Level Wargaming*, Alexandria, Va.: Center for Naval Analyses, September 2004, pp. 1–2.

fight" as a meaningful element of the game adjudication. Commercial products fared better in that more systems had a morale characteristic or similar aspect of play. This includes such behaviors as delayed or missing orders, eroded performance, suppression, hesitation, and pinning, as well as surrendering or retreating. However, in many cases the representation of these mental and morale factors did not align with findings from the empirical literature. For example, many games leaned on simplistic cultural stereotypes (e.g., the French always surrender but the Japanese fight to the bitter end).[28] In other words, even when mental and morale factors are included, they are not included well.

In informal discussions with designers, we identified three basic approaches to including the mental and morale spheres in games. The first approach is to rely on players to model mental and morale changes in the force. For example, a game would model deception by presenting partial information to players, monitoring their decisionmaking process, and recording their emotional state as additional information is revealed. In many ways, this is an attractive notion that draws on what makes games unique as a form of analysis: that actual humans make actual decisions during the game. For example, games that focus on C2 can leverage this fact to experiment and clarify the effects of information on decisionmaking.[29]

However, this approach comes with several fundamental flaws. The first is that capturing players' mental models and decisionmaking processes in a way that is transparent to consumers of analysis (such as commanders and force planners) is challenging even in the best-designed games. If the mental and morale spheres are captured poorly, it will be difficult to consider these factors in analyses. Game results may ignore the mental or morale impact on forces or may misconstruct player models, contributing to misleading results. Another result of poor documentation is that it becomes difficult for scholarly discourse to grow up around these issues. As a result, practice in this area of gaming does not have an opportunity to improve over time.

Second, it is not always possible or ethical to accurately simulate psychological and emotional effects in the artificial environment of a wargame. For example, a MISO effort designed to affect behavior by invoking fear through danger to loved ones would be unethical to represent in a way that credibly invoked fear for the safety of players' family. Even less evidently unethical situations, such as staff decisionmaking processes, may raise real questions about assumptions that players' emotions and decisions in the low-risk, time-compressed setting of a game can mimic real-world conditions. It is difficult to create the stresses associated with combat and command decisionmaking in battle. At the very least, designers who opt for this approach must devote far more

---

[28] Connable et al., 2018, pp 113–156.

[29] Rubel, 2003; Hank J. Brightman and Melissa K. Dewey, "Trends in Modern War Gaming: The Art of Conversation," *Naval War College Review*, Vol. 64, No. 1, 2014, Article 4.

attention to defending the credibility of assumptions regarding player reactions rather than assuming their analytic value.

Another basic approach to including the mental and morale spheres in games is to consider mental and morale factors as one part of broader rules for establishing behavior. For example, the development of unit breakpoints (discussed in more detail in Chapter Six) often consider the effects of casualties on morale and purely physical attrition. Again, the problem here is that conditions in the mental and morale are often opaque to players and consumers of game analysis. Final documentation might describe what the breakpoint is, but rarely does it indicate what factors designers were considered to generate it. Again, to some extent, this approach can be improved simply by better documenting game rules so that it is more apparent how various factors come into play. However, even when they are transparently documented, mental and morale factors are too often treated only as contributing factors and thus they take on secondary roles from the perspectives of players and analysts. For example, an adjudication model might allow ratings of unit training, esprit de corps, and leadership quality to adjust how many causalities forces must sustain to be rendered combat ineffective.[30] However, players do not interact with this information directly in the game because it is marginal to the core dynamics of game play and cannot be changed by player actions, so it may not have much impact on players' decisions. In effect, this style of game underestimates the impact of the mental and morale spheres, producing systematically biased results.

Instead, we argue for a third approach: As a default design choice that mental and morale factors are as prominent as physical phenomena in game design. For example, if players receive information about unit attrition over time, they should also be given information about the quality of decisionmaking and unit will to fight. If players are asked to describe the intended effects of an attack on enemy forces, that description should include the impact on information that will shape all three spheres of war. Objectives, and the assessment of whether those objectives are achieved, should adjudicate activities in all three areas. Of course, the exact prominence of battlefield phenomena will be driven by the game's objectives, but moving the default closer to explicit modeling of all three spheres is critical for OIE to get a fair shake in games.

A major barrier to this recommendation is the absence of well-developed and generally accepted abstract models of the mental and morale spheres that can be adopted in game design. Force-on-force combat will change in response to relative capabilities and terrain, but there is a wide variety of well-accepted tools for modeling these phenomena. In contrast, existing conceptual models of such phenomena as will to fight and information operations (IO) are often designed to be portable, meaning that they

---

[30] Such models are quite common in commercial wargames, and designers may find an examination of these games enlightening (see Connable et al., 2018, chapt. 3). However, it is critical for professional designers to remember that commercial game models are designed for enjoyable play rather than empirical rigor and should rarely be adopted whole cloth for analytic purposes without careful study.

capture relevant factors while recognizing that their relative importance and the relationships between them must be tailored to a specific context.[31] Although this may require additional research on the part of game designers, it is feasible to make context-specific adjustments if such an effort is prioritized. The other criticism is that, often, these models are seen as less credible than better-established kinetic models and thus introduce a risk of misleading players and consumers of analysis. It is certainly true that game designs should be transparently documented so skeptics can see how the model was implemented, but failing to represent mental and morale phenomena introduces immediate, profound bias. By including even imperfect models of the mental and morale spheres, games can open up space for their explicit consideration, allowing debate, discussion, and learning.

The need to consider mental and morale factors in combat should pervade game design—from setting objectives for players to considering what information is accessible to them and designing adjudication functions. For example, carefully crafted victory conditions might track and specify national will to fight as a key characteristic that is affected by game events, and its decline to a certain level might signal defeat for one side, even if it has strong and willing formations in positions of relative advantage on the physical battlefield. Similarly, explicit modeling of what information is available to what commanders should shape information available to players, and what portions of their orders are implemented.[32] Finally, assessments of the mental and morale states of subordinate units should affect which orders are followed and how well. Only by consistently considering the mental and morale spheres throughout games is it possible to capture the full importance of all three spheres.

---

[31] See Connable et al., 2018, pp. 33–35 for one such discussion.

[32] Perla et al., 2004.

CHAPTER FIVE

Defeat Mechanisms

Accepting that to capture the true nature of war, wargames must capture the mental and morale spheres (and that the mental and morale spheres predominantly inhabit the IE and can be affected by OIE), we now turn to the question of how this can best be achieved. As is true for anyone who develops abstract models of complex phenomena, wargame designers must make choices about what aspects of conflict to abstract or omit from the game to create something simple enough to be analytically useful. In considering what aspects of the mental, morale, and physical should be included in games, we must examine what aspects of conflict are key and why.

To begin to build a bridge from the abstract notion of defeat across the three spheres to the concrete actions and supporting mechanics needed to build such an operational game,[1] we return to military theory and consider what are described as *defeat mechanisms*, the fundamental logic behind the various methods that a force can use to prevail over enemy opposition (including actions in and through the IE, which can affect all three spheres but affect the mental and morale most directly). Because defeat is the primary objective of military campaigns, it makes sense that the implicit objective given to players in operational wargames is to defeat adversaries. If OIE can meaningfully affect defeat outcomes, then players will be more likely to employ such capabilities during games.

We first review the literature on defeat mechanisms, identify a range of ways beyond attrition that forces can defeat one another, and describe how these various mechanisms draw on the mental, morale, and physical spheres, as well as how OIE or the IRCs might contribute to each mechanism. We then discuss how these defeat mechanisms could be depicted through game mechanics as a prompt for designers to begin considering this topic in their work.

[1] We recognize that there are many other types of games conducted by the U.S. military.

Defeat Mechanisms Identified

Defeat is a core objective of warfare and a combination of effects in the mental, morale, and physical spheres. Our review of military theory revealed a range of defeat mechanisms beyond simple destruction or attrition, including annihilation, dislocation, exhaustion, disruption, disintegration, preemption, isolation, circumvention, forestalling, and compellence/impellence/deterrence. We discuss each in turn.

Destruction

Destruction is a the simplest and most straightforward of defeat mechanisms and has an inescapable logic. Destruction applies combat power on an enemy capability to damage it such that it cannot perform any function.[2] A foe whose capabilities have been destroyed can no longer do anything at all, is no longer a threat, and is certainly and unequivocally defeated. Some have argued that destruction is not actually a defeat mechanism: An enemy is not necessarily defeated until it is completely destroyed, and if it is defeated to a point short of complete destruction, then the mechanism of that defeat was not destruction.[3] Others have argued (following Clausewitz) that destroying a force simply means imposing a condition under which it can no longer carry on fighting, which might be achieved through physical destruction, but might also be achieved by destroying the force's moral cohesion, a topic we return to later.[4] Destruction as a defeat mechanism is also called *attrition*, which implies wearing down an adversary's capacity to fight faster than it can replace or reconstitute that capacity.[5]

Annihilation

Annihilation is a form of destruction in which one seeks to reduce an adversary's capability all at once, in a single decisive battle or a series of shorter battles.[6] The goal is to destroy the enemy's ability to fight as a cohesive force and sue for peace. This differs from attrition in that it is not about grinding down adversary forces faster than they can be replaced, and it differs from destruction in that it implies that a determination to shatter the will of the enemy—both the will of the annihilated forces and the will of the nation providing them.

Dislocation

Dislocation is also called the *indirect approach* and is the foundational defeat mechanism of maneuver warfare. Dislocation is about applying strength to an opponent's

---

2 Army Doctrine Publication 3-0, *Operations*, Washington, D.C., July 2019, p. 2-4.

3 Leonhard, 1998, p. 211.

4 Friedman, 2017, p. 17.

5 Echevarria, 2017.

6 Echevarria, 2017.

weakness and finding ways to unbalance the foe and thus negate any remaining strengths or render them irrelevant.[7] Surprise of one form or another is central to many efforts at dislocation. Some theorists suggest that dislocation is not directly a defeat mechanism but that successful dislocation exposes the enemy to the effects of other defeat mechanisms.[8] Leonhard identifies four types of dislocation: positional (in which enemy strength is irrelevant because it is in the wrong place or facing the wrong way), functional (where technology or combined-arms threats negate enemy strength), temporal (use of surprise or extreme tempo), morale (offsetting the enemy's strength by defeating its will).[9]

Exhaustion

Exhaustion is like the psychological version of attrition, grinding away at a foe's willingness to continue to fight.[10] Exhaustion as a mechanism can be employed to affect enemy forces, reducing their morale or willingness to participate in combat, but it can also affect the public confidence of a nation or threaten its political balance of power. Defeat occurs when soldiers are no longer willing to fight or when public support for a conflict wanes to a sufficient extent. Like attrition, exhaustion should be presumed to take considerable time and be the consequence of accumulated effects.[11]

Disruption

Disruption occurs when a military formation exists but is unable to function; paralysis is a form of disruption, or dislocation might result in disruption.[12] Disruption might be brought about through firepower, maneuver, surprise, or other factors or effects.

Disintegration

Disintegration occurs when the will and the moral cohesion of enemy forces have been broken. Traditionally, attempts to employ this mechanism would rely on closely coordinated shocks through fires and ground assaults; without close coordination and strong effects from both, the attempt to disintegrate can just become an application of

---

[7] Echevarria, 2017.

[8] Army Field Manual 3-07, *Stability*, Washington, D.C., June 2, 2014, p. 4-8.

[9] Leonhard, 1998, p. 64.

[10] Echevarria, 2017.

[11] Friedman, 2017.

[12] Eado Hecht, "Defeat Mechanisms: The Rationale Behind the Strategy," *Infinity Journal*, Vol. 4, No. 2, Fall 2014; Lamar L. Tooke, "Blending Maneuver and Attrition," *Military Review*, Vol. 80, No. 2, March–April 2000.

attrition.[13] One could imagine that other means could contribute to disintegration as well, such as efforts that lead to alienation or desperation among the targeted force.[14]

Preemption

Preemption is all about speed, either surprising the enemy and crushing it or using speed to decrease the number of options available to the enemy. Preemption is initiating an action before the enemy is ready. Preemption forces foes to react rather than proceed with their own plans, and it is another way to attack the moral cohesion of a foe.[15] Extrapolated as an independent defeat mechanism rather than something that enables or leads to another defeat mechanism, preemption could be the seizure or accomplishment of a decisive objective before the enemy mount an opposition.

Isolation

Isolation occurs when all or part of an enemy force is cut off from needed communication and support or unable to affect a friendly force. "Isolation involves marginalizing the enemy's critical capabilities or limiting the enemy's ability to influence relevant events."[16] Isolation could result from an extended lure out of position, stranding a force on the wrong side of terrain (such as a river that it lacks the means to cross), or penetrating enemy lines and cutting critical lines of communication.

Circumvention

Circumvention is a subset of isolation in which the enemy is actually surrounded rather than isolated by other means.[17] Although circumvention can also mean avoidance rather than encirclement, avoidance as a defeat mechanism would better fit under dislocation, isolation, or exhaustion, depending on the specific circumstances.

Forestalling

Forestalling is the effective application of security.[18] To forestall enemy forces is to prevent them from achieving their objectives. Denied long enough, with sufficient consistency, or with sufficient violence, enemy forces would eventually succumb to one of the other defeat mechanisms.

---

[13] Douglas J. DeLancey, *Adopting the Brigadier General (Retired) Huba Wass de Czege Model of Defeat Mechanisms Based on Historical Evidence and Current Need*, Fort Leavenworth, Kan.: School of Advanced Military Studies, U.S. Army Command and General Staff College, 2000.

[14] Bruce Watson, *When Soldiers Quit: Studies in Military Disintegration*, Westport, Conn.: Praeger, 1997.

[15] Tooke, 2000.

[16] Army Field Manual 3-07, 2014, p. 4-8.

[17] Hecht, 2014.

[18] Leonhard, 1998.

Compellence, Impellence, and Deterrence

Compellence, impellence, and deterrence are not included in many theorists' lists of defeat mechanisms. However, if we accept that warfare is a contest of wills and recognize virtue in not just crushing an opponent's will but in also imposing one's own will, then this should be considered a defeat mechanism. Getting an adversary to do (or not do) what one wants is also a way to achieve tactical, operational, and strategic objectives.

Connecting the Defeat Mechanisms to the Spheres of Conflict

Although each of these defeat mechanisms has its own logic, some categorization is possible. For example, preemption, dislocation, and disruption are the cornerstones of the maneuverist philosophy of warfare. Similarly, attrition is focused primarily on physical effects, while disintegration focuses on morale effects, and dislocation (arguably) emphasizes effects in the mental sphere.[19] Table 5.1 lists each of the defeat mechanisms and notes the action of each in relation to the three spheres of warfare: mental, morale, and physical. The table highlights the primary effects of each defeat mechanism. The rightmost column of Table 5.1 considers the possible role for OIE and the IRCs in contributing to each mechanism. Understanding which logics of defeat contribute to effects in which of the spheres (and how), as well as where and how OIE might contribute to those logics, constitutes an important step toward incorporating these logics (and OIE) into wargames.

Table 5.1
Defeat Mechanisms and Their Relationships to the Mental, Morale, and Physical Spheres of War

Mechanism	Mental Sphere: Perception, Judgment, Decisions	Morale Sphere: Morale, Will, Leadership	Physical Sphere: Capability	Possible Role of OIE/IRCs
Destruction/ attrition	Destruction of C4ISR prevents decisions from being made or conveyed.	Morale and will are harmed as a secondary effect of damage; sufficient destruction will break a foe.	Physical sphere is the target of destruction; a destroyed foe does not oppose one's ends.	Amplify the perception and shock of destruction or contribute to C4ISR damage.
Annihilation	C4ISR is significantly disrupted by overwhelming force.	Shatter and overwhelm a foe's will to fight and support further conflict at the national level.	Devastating destruction significantly weakens a foe's capabilities.	Amplify perception and shock of devastation.

[19] DeLancey, 2000.

Table 5.1—Continued

Mechanism	Mental Sphere: Perception, Judgment, Decisions	Morale Sphere: Morale, Will, Leadership	Physical Sphere: Capability	Possible Role of OIE/IRCs
Dislocation	Render enemy plans and options irrelevant, limiting the ability to respond effectively.	Contribute to confusion and weaken morale and moral cohesion.	Make a foe's capability irrelevant; even if it is intact, it is not where it needs to be to mount an effective opposition.	Constitute the core of OIE: C2W, stratagem, influence, deny, and deceive.
Circumvention/ surrounding		Being surrounded is tough on will and moral cohesion.	Isolate foe from supply and support capabilities.	Magnify the foe's perceptions of the vulnerability of its position.
Exhaustion	Declining will leads to decision hesitancy and risk aversion.	Grind away at will to fight at the tactical, operational, and strategic/ national levels.	Physical damage may be required, but exhaustion affects physical capabilities only indirectly.	Support and amplify perceptions of unacceptable costs relative to prospects for success.
Disruption	Disrupt C4ISR.	Act to shock, surprise, and suppress.		Constitute the core of OIE: C2W, stratagem, influence, deny, and deceive.
Isolation	Prevent connection with higher C4ISR.	Impose paralysis and hopelessness by cutting off enemy forces from C2 and support.	Isolate foe from supply and support capabilities.	Use information warfare, destroy/ usurp C4ISR, and magnify the psychological effects of physical force.
Disintegration		Attack coherence and cohesion and troops' ability to respond to C2, including through surprise and shock effects.	A certain amount of violence and noise are required, even if actual capability degradation is minimal.	Destroy or usurp C4ISR and magnify the psychological effects of physical actions.
Preemption	Plans disrupted, the foe is forced into a reactive mode	Act to cause disappointment, despair, and surprise.		OPSEC is essential, with MILDEC and SIGMAN playing a role.
Forestalling	It is difficult for the enemy to move to the next stage in its plans if it is prevented from accomplishing the first.	Undermine will to continually fail/lose.	Foe expends forces or capabilities without making progress.	Defensive contributions protect friendly C4ISR and morale.
Compellence/ impellence/ deterrence	Change the decision calculus changed and decisions made as a result.	With enemy will suborned, one's own can be imposed.	No primary effect on capability.	Constitute the core of OIE: C2W, stratagem, influence, deny, and deceive.

NOTE: Shading indicates primary effects from each defeat mechanism. C4ISR = command, control, communications, computers, intelligence, surveillance, and reconnaissance.

Similar to Table 5.1, Table 5.2 lists the defeat mechanisms discussed in this chapter but also broadly gauges the extent of each mechanism's contribution in each sphere. The rightmost column grades the extent of the possible IRC contributions. The scale consists of three levels, with a single plus (+) denoting some contribution or impact, a double plus (++) denoting a significant contribution or impact, and a triple plus (+++) denoting a major contribution or impact. The scoring in the table is preliminary, based on the authors' judgment, and could change under different sets of circumstances or assumptions.

Together, Tables 5.1 and 5.2 communicate several points. First, there are ten defeat mechanisms other than attrition. Second, all defeat mechanisms—even attrition—have an impact beyond physical effects, in the mental sphere, the morale sphere, or both. Third, IRCs (and, thus, the IE) have a potential role in all defeat mechanisms, with the greatest potential role in dislocation, disruption, preemption, and compellence/impellence/deterrence. Considering the three spheres and multiple defeat mechanisms opens up combat adjudication to effects from OIE while still capturing the contribution of traditional physical destruction and damage.

Table 5.2
Defeat Mechanisms and Their Potential Impact in the Mental, Morale, and Physical Spheres

Mechanism	Mental Sphere: Perception, Judgment, Decisions	Morale Sphere: Morale, Will, Leadership	Physical Sphere: Capability	Role of IRCs
Destruction/ attrition	+	+	++	+
Annihilation	++	+++	++	+
Dislocation	++	++	+	+++
Circumvention/ surrounding		++	+	+
Exhaustion	+	++		+
Disruption	++	++		++
Isolation	++	++	+	+
Disintegration		+++		+
Preemption	+++	+		++
Forestalling	+	+	+	+
Compellence/ impellence/ deterrence	++	++		+++

NOTE: + = some contribution/impact. ++ = significant contribution/impact. +++ = major contribution or impact.

Game Mechanics for Defeat Mechanisms

Having described the nature of these eleven defeat mechanisms, we now discuss how they can be represented in operational military wargames. Four primary areas of design are important in this context. First, we discuss the prompts provided to players that lay out game objectives. Whether these prompts are formal or informal, the language used to motivate game play should enhance understanding of what drives player choices. Second, we consider the issue of player situational awareness—that is, what information players can access about the state of the environment and other actors to inform their decisions. Third, we consider the actions that are made available to players in the game. Finally, we review the types of adjudication tools needed to represent key phenomena. In all cases, the goal is to present new options for a game designer to consider employing in a range of operational games, rather than to recommend a fully fleshed-out game design.

The objectives that guide player choices in operational games are an ideal place to widen the aperture of options for players to consider as they attempt to defeat opponents. Often, adversary objectives are framed offensively (e.g., invade and secure *x* territory), while U.S. military objectives seek to restore the status quo while defending U.S. and allied interests in a given theater. Such broad objectives leave a great deal of room for players to determine how best to prosecute the operation. Similarly, specifying a range of possible victory conditions (e.g., all enemy forces eliminated or captured, or all enemy forces withdraw to within their own borders) might prompt players to think about different approaches to defeating their foe. Furthermore, simply offering multiple potential mechanisms for players to consider may prompt a richer, more explicit discussion of options. For example, prompting players to consider methods to dislocate, circumvent, or isolate forces may spark explicit conversations about strategic geography and its potential effects on operations rather than encouraging players to default to a strategy of enemy destruction or exhaustion. Regardless, objectives should always be framed to provide meaningful choice to players rather than dictating their concept of operation.

Next, a designer must consider how to represent the environment and information about player capabilities. This includes both what data should be available in the game and who should have access to which types of data. Most important is ensuring that all three spheres of war are tracked during the game. Many of the common catalogue game design elements—for example, geographic maps or counters indicating numbers, capabilities, and locations of troops—are geared toward the physical. However, there are many options for introducing trackers, unit capability indicators, and nongeographic maps that offer ways of representing the mental and morale spheres as well. For example, displaying both "will to fight" and casualty counts for each unit can make players more conscious of the morale sphere when making decisions. As forces fight and face the demoralizing effects of lost equipment, casualties, enemy messaging, and poor leadership, will to fight and other markers of combat effectiveness can be degraded. Conversely, the effects of training, good leadership, and moral conviction can promote high levels of will to fight and the potential for heroic action. Players can then incorporate this information when making decisions. Should injured forces with low will to fight be prioritized for evacuation to limit the risk of surrender? Where should forces with strong will be positioned to best secure objectives? In the same vein, information about the quality of communications between different units at any particular point in the game can make the mental sphere more prominent. In effect, visualization becomes a way to counter "out of sight is out of mind" planning by players.

One critical tension to consider is that if information about the mental and morale spheres is merely added to an existing interface for physically focused game play, the amount of information players are required to process can be unwieldy. Generally, operational games ask players to speed up planning processes, playing though days' worth of time in mere hours. To make this viable, the game uses abstraction and simplification to reduce the amount of information presented. To maintain the same level of overall complexity, it may be necessary to reduce the amount of information players receive about the physical sphere to devote more time and attention to the mental and morale spheres. These choices must be made carefully, in line with the game's analytic objectives, or risk the game failing to meet its goals.

At the same time, robust play of events in the mental sphere often requires that players work with incomplete information. To simplify the logistics of play, games often operate with near-perfect information available to all sides. However, even early Prussian staff games recognized the need for differentiated information to mimic the limits of information available to each side. Although it is not always necessary, strategies that depend on surprise and deception do require an ability to track different levels of situational awareness, something we discuss in more detail in Chapter Six. Again, hidden information often increases the complexity of the game for players and adjudicators, so thoughtful consideration of where it is needed and what other aspects of the game can be simplified is critical for maintaining playability.

The third area of design that is important for representing defeat mechanisms concerns the actions available to players in the mental and morale spheres. For example, players may well assume that they do not have the option to surrender or withdraw forces unless the option is explicitly provided. Robust play of OIE, in which players understand the choices available and their potential benefits, is also critical. Providing the necessary context for the decisionmaking may require player education either before the game or through the use of in-game mentors.

For information and actions to be meaningful, the mental and morale spheres must have an impact on the course of the battle, requiring adjudication of effects in these spheres. If players are provided with information about the mental and morale spheres of the game and are able to act in these realms but do not see meaningful consequences from those choices, these spheres will rapidly fade from player attention. This often occurs when OIE are "bolted" onto an existing game: Players may be asked to complete IO plans and may opt to deploy one set of messages over another, but when OIE do not have an impact on the battlefield, players do not see a reward to continuing to apply these tools and will focus scarce attention on other areas. This means that adjudication systems must be designed so that actions in the mental and morale spheres affect the results of combat or other relevant game outcomes, and these effects must be transparent and defensible to players.

Perhaps the most fundamental shift required is that subordinate forces will not always follow the orders of the players, since subordinate commanders and forces may be unable or unwilling to follow orders. As a general rule, wargames are designed so that subordinate forces follow orders as given. This design choice has been defended on analytical grounds, since players and analysts need to develop a reasonably clear narrative about cause and effect to communicate the drivers of game results. However, as discussed earlier, this challenge can be mitigated through greater transparency in adjudication techniques. The more difficult barrier to adopting this style of design is that, when forces deviate from player decisions, it reduces players' agency and, potentially, their engagement and enjoyment. However, this concern should not stop designers from implementing such rules, both because hobby games provide a wealth of examples of engaging games that involved nonvoluntary actions and because the research objectives of an analytical game should outweigh concern with player enjoyment.

It's also important to note that the choice to explicitly include the mental and morale spheres in adjudication need not dictate the style or tools of adjudication. Often, rigid adjudication systems are helpful in making rules more transparent, simply because they are documented. However, approaches based on expert judgment rather than preformulated models can also fulfill this need. For example, in a classic expert-adjudicated seminar-style game, mental and morale considerations can be included by prompting the expert adjudicator to consider those factors. How would subordinate and senior commanders make decisions? How might the psychological impact of a message determine soldiers' willingness to fight? Would the public still support the war, given the outcomes of battle? In some cases, simply changing how experts are prompted may be enough. In others, it may make more sense to incorporate additional types of expertise. For example, an adjudicator with combat experience might be better able to speak to the mental and morale effects of combat then a weapon system expert. Social scientists can also shed light on national, bureaucratic, small-group, and personal decisionmaking.

In addition to these cross-cutting considerations, adjudication of specific defeat mechanisms may require specialized tools. Table 5.3 offers some examples of game design characteristics needed to support adjudication. These suggestions are predicated on the assumption that some or all of the recommendations outlined in this chapter are adopted, particularly those related to explicitly tracking will to fight and hidden information. All may not be necessary or appropriate in every game, but, again, the intent is to provide a sense of the possible to inform future designs.

Table 5.3
Sample Game Mechanics to Adjudicate Defeat Mechanisms in the Mental, Morale, Mental, and Physical Spheres

Mechanism	Mental Sphere: Perception, Judgment, Decisionmaking	Morale Sphere: Morale, Will, Leadership	Physical Sphere: Capability
Destruction/ attrition/ annihilation	For subordinate units, set probabilistic breakpoints that consider a range of mental, morale, and physical factors.	As causalities are adjudicated from force-on-force engagements, the measure of a unit's will to fight is reduced for remaining forces.	Design meaningful degradation steps for each unit type. Consider how unit capabilities decline in response to attrition.
Dislocation and circumvention	Set time requirements to change orders to ensure that units cannot respond artificially quickly to changing conditions.	Establish rules for shock effect and surprise, as well as changes to the breakpoint.	Impose physical consequences for units attacked on flanks or by surprise.
Exhaustion	Forces for nations nearing exhaustion may be more risk-averse, inclined to both avoid unnecessary losses in a losing effort and avoid results that further contribute to domestic war-weariness.	Track national will to fight and factors that reduce it. Units may face declining morale alongside decline in domestic support for the mission.	As national will to fight is exhausted, reinforcements and supplies may be reduced or operational constraints may be imposed.
Disruption, isolation, and surrounding	Track lines of communication. Units that are not "in command" assume default behavior (e.g., following last available order, responding in an adjudicator-determined manner to conditions, doing nothing).	Track lines of communication. Units that are not "in command" receive a will-to-fight penalty.	Track key unit supplies and expenditures. Units low on supplies are physically hindered (e.g., no food) or cannot act (e.g., no ammunition).
Disintegration	Subordinate units may be modeled as more likely to fail to respond to orders.	Low will to fight can result in unit breakpoint behavior, under which orders cannot be carried out.	
Preemption	Natural player confusion can be heightened by restricting time available to design moves in response to a surprise attack. Subordinate units may be modeled as more likely to fail to respond to orders.	Surprise attacks create a will-to-fight penalty.	
Forestalling	Stalled forces cannot accept new orders until they successfully disengage.	Stalled forces take a will-to-fight penalty.	Unit cannot physically move until winning an engagement. (These mechanics are known as *zones of control* in commercial games.)
Compellence/ impellence/ deterrence	Simulated higher-level commanders who can terminate the war or change objectives based on outcomes.		

Related Topics

Two topics related to this discussion of defeat mechanisms are breakpoints—points at which forces are considered combat-ineffective—and surprise. In the next chapter, we discuss breakpoints and surprise in greater detail, describing opportunities for the more dynamic inclusion of breakpoints in wargames and some considerations for deploying hidden information when wargaming surprise.

CHAPTER SIX

Breakpoint and Surprise

This chapter offers some suggestions for incorporating breakpoints into games as a way of better representing the mental and morale aspects of combat in adjudicating combat operations. The chapter begins with a discussion of the theory behind breakpoints based on past work in gaming, modeling, and simulations. It then suggests several practical approaches to game design with breakpoints consistent with the principles we lay out in this report. The discussion then gives similar treatment to the topic of surprise.

Theory of Breakpoints

Traditionally, units are weakened or removed from play in wargames based exclusively on attrition-based thresholds. In some wargames and simulations, this is called the *breakpoint*, or "forced changed in combat posture"—the point at which a unit has suffered sufficient casualties that it becomes "broken" or combat-ineffective and surrenders or routs.[1] The idea of a breakpoint has the virtue of acknowledging defeat at a point short of the complete destruction of a formation, but, as usually implemented, it has a range of shortfalls.

First, the traditional approach to breakpoints ties defeat solely to casualties rather than to the harder-to-quantify but more essential will to fight. Research suggests that attrition is not what drives fundamental changes in adversary behavior. For example, a study of 80 battles from World War II and the Arab-Israeli wars identified the distribution of "reasons for abandoning an attack or defense," shown in Table 6.1. Casualties represented only 12 percent of breaks, whereas the vast majority were caused by alternative defeat mechanisms identified in Chapter Five.[2] Without making claims

[1] Robert L. Helmbold, *Decision in Battle: Breakpoint Hypotheses and Engagement Termination Data*, Santa Monica, Calif.: RAND Corporation, R-772-PR, 1971.

[2] T. N Dupuy, "A New Approach to a Breakpoint Methodology," draft article, February 19, 1987, cited in Dan J. McConnell, *Investigating a Coherent Framework for Deception Analysis*, Monterey, Calif.: Naval Postgraduate School, June 1988.

Table 6.1
Reasons for Abandoning an Attack or Defense

Reason for Break	Frequency
Threatening enemy maneuver (i.e., envelopment or penetration)	39%
Withdrawal of adjacent unit	13%
Reserves exhausted	12%
Casualties incurred	12%
Reaction to surprise	8%
Truce or general surrender	6%
Other, including logistics, planned withdrawal, and weather	10%

SOURCE: Dupuy, 1987, cited in McConnell, 1988.

NOTE: The table shows frequency across a data set of 80 battles from World War II and the Arab-Israeli wars.

that this distribution of results is generalizable, the research certainly demonstrates the importance of alternative defeat mechanisms in driving breaks.

The notion of a breakpoint fails to fully respect that "[d]efeat is a psychological state" or that "defeat is 90 percent moral in nature."[3] Existing work offers a range of alternative factors to consider.[4] The following are some examples:

- force strength and attrition, with subfactors including
  - unit strength
  - force and combat power ratio
  - perception of relative force size
  - artillery and air attack
  - causalities
  - casualty rate
  - loss of equipment
- tactics and maneuver, with subfactors including
  - tactical plan
  - relative posture
  - enemy maneuvers, including envelopment, encirclement, and penetration
  - forces in vulnerable position
  - ineffective friendly maneuver, adjacent friendly units withdrawing, or failure of adjacent units to advance

3 Storr, 2009, p. 52; Leonhard, 1998, p. 211.

4 See, for example, Janice B. Fain, Richard C. Anderson, Trevor N. Dupuy, Gay M. Hammerman, and Charles F. Hawkins, *Forced Changes of Combat Posture*, Fairfax, Va.: Historical Evaluation and Research Organization, Data Memory Systems Inc., September 30, 1988, p. V4-6.

  - low troop-to-frontage ratio
  - hasty unit commitment on new ground
  - attacker's rate of advance
- resources and current status of forces, with subfactors including
  - lack of fire support (including artillery and airpower)
  - lack of reinforcements and reserves
  - troop exhaustion
  - supply shortage, particularly ammunition, food, and water
  - lack of communication or confusion over orders
  - intelligence failure and surprise
  - enemy reinforcements
  - poor staff work
- pre-engagement status of forces, with subfactors including
  - condition of troops at the beginning of engagement
  - duration of combat
  - lack of combat experience
  - training
  - previous training or experience similar to current situation
  - time in line before engagement
  - poor physical fitness
  - success in last combat engagement
  - fatigue
  - motivation
  - morale and esprit de corps
  - expectations regarding assignment
  - number of new personnel replacements
  - leadership
  - poor staff work
  - confusion over orders
  - poor maps
- mission, including the imperative of the assigned task
- physical environment, with subfactors including
  - unusually stressful or rugged terrain
  - previous experience in the area or terrain type and climate
  - changes in weather
  - poor roads
- exogenous factors, including
  - truce
  - general surrender
  - orders to withdraw.

These factors reinforce the same basic point—that causalities are a subfactor rather than a primary driver of units breaking.

Second, too often, breakpoints are implemented deterministically: A unit hits a particular, known casualty threshold and ceases to operate. This approach has long been critiqued, with scholars and modelers arguing that it would be more appropriate to model unit's tolerances as a probabilistic curve, given the complexity of the interactions between different factors and high degree of uncertainty in any commander's ability to correctly predict when forces will break.[5]

Finally, an examination of the empirical literature illustrated that not all forces break in the same way. Table 6.2 summarizes past RAND work laying out a range of different behaviors that can be driven by the will to fight.

Such analysis shows a much wider range of potential behaviors that would affect combat outcomes then simply leaving the battlefield. Thinking about these behaviors as a continuum, rather than simply as a binary set of states (fully obedient or fully disobedient), is key to capturing actual combat dynamics.

Table 6.2
Possible Combat Behaviors Resulting from Decisions Driven by Will to Fight

Factor	Description	Type
Heroism	Hyperaggressive individual behavior that can inspire other soldiers	Positive
Aggression	Attacking or defending vigorously; help degrade adversary will to fight	Positive
Competent assault	Perform offensive mission with a calm, workmanlike attitude	Acceptable
Competent defense	Perform defensive mission with a calm, workmanlike attitude	Acceptable
Hesitation	Delay in following orders or taking action	Harmful
Pinned	Unwillingness to move under fire but may return fire	Harmful
Freezing	Unwillingness to act and a descent into incapacitation	Harmful
Disobedience	Refusal to follow a combat order	Harmful
Panic	Soldier allows fear to dominate resulting in ineffective behavior	Harmful
Rout/flee/break	Running away from combat	Harmful
Surrender	Quit fighting and submit to enemy control	Harmful

SOURCE: Connable et al., 2018, p. 121, Table 3.4.

[5] Helmbold, 1971.

Wargaming Breakpoints

These theoretical findings suggest several critical improvements to how breakpoints are designed into games. Empirical evidence suggests that they should be built around a broad understanding of the morale sphere or will to fight, rather than focused on causalities.

However, there is still a considerable range of options for implementing a system to adjudicate breakpoints that follows these precepts. For example, in a seminar-style game, a designer might opt for a matrix adjudication approach. In matrix games, players state their action, describe the outcome they think they can get, and explain why the action will be successful. Other players then get to weigh in on why the action will not be successful. Expert adjudicators can then determine the outcome based on their judgment, or, more often, assign a likelihood of each possible outcome and determine the outcome using dice or another randomizer.[6] The following notional dialogue reflects how such an approach might proceed in practice:

> Player A: I'm going to attack Player B's formation here, and Player B's forces are going to break and run. The reason I think they will do this is that they are demoralized from bad leadership: Our leaflets are effective at encouraging deserters, and Player B's troops need to return to their villages to deposit their pay at this time of the month anyway because there is no modern banking system in this part of the country.
>
> Player B might weigh in: Player A is not going to be successful because my troops all come from the same village and would lose face if they ran. They feel safe behind the fortifications they've made since the last attack. Morale is high because of [a high-level commander's] recent visit. Plus, many of the enlisted personnel are illiterate so can't read your leaflets anyway.
>
> [Dialogue continues along these lines.]
>
> After arguments are presented, the adjudicator might have Player A roll a die and award that player partial success: The troops in one position broke and ran but those in another position had the high ground and more ammunition and dug in. Some of the troops from the first position joined those in the second position, so the second position is now more heavily defended than before.

This approach includes explicit conversation about a range of potential mental and morale factors, is probabilistic, and allows for a range of behaviors and outcomes. However, it does not require a rigidly defined model to be in place at the start of game

---

6 For more on this approach, see John Curry and Tim Price, *Matrix Games for Modern Wargaming: Developments in Professional and Educational Wargaming*, History of Wargaming Project, 2014.

play. As a result, such an approach may be particularly useful when examining poorly understood phenomena for which the designer cannot build a credible rigid adjudication system, or where a more discursive, qualitative manner of play produces information that better supports post-game analysis.

An alternative is to adopt the traditional rigid adjudication system more typical of past breakpoint modeling efforts but to reformulate it to better follow the principles outlined here. In this case, it is important to remember that a model that tries to capture everything will not be clear enough to be useful. Therefore, the goal of the designer should be to select factors and subfactors to assemble a tractable model that captures key dynamics for the context of interest.

One potential approach is for the designer to build a fairly simple breakpoint adjudication system focused on assessing each unit's will to fight on a five-point scale. At the start of conflict, will to fight might depend on several factors: level of training, the experience of force, and time in the field, for example. Each level of will to fight comes with a different probability of the unit breaking: Units with high levels of will to fight would have a very low probability of breaking, where units with a low will to fight would have a much higher chance of breaking, as illustrated in Table 6.3. At the beginning of each turn, a commander would use a dice roll to determine whether that unit would follow orders.

Such a system could be further refined, both by adding dice rolls modified to represent an increase or decrease in the probability of a unit breaking based on the context

Table 6.3
Potential Breakpoint Outcomes, by Unit Will to Fight

Results of Rolling Two Six-Sided Dice	Level 1 Poor Will to Fight	Level 2	Level 3 Moderate Will to Fight	Level 4	Level 5 High Will to Fight
12	Acts on order	Acts on order	Acts on order	Acts on order	Acts on order
11	Acts on order	Acts on order	Acts on order	Acts on order	Acts on order
10	Acts on order	Acts on order	Acts on order	Acts on order	Acts on order
9	Acts on order	Acts on order	Acts on order	Acts on order	Acts on order
8	Acts on order	Acts on order	Acts on order	Acts on order	Acts on order
7	Acts on order	Acts on order	Acts on order	Acts on order	Acts on order
6	Hesitation	Acts on order	Acts on order	Acts on order	Acts on order
5	Hesitation	Hesitation	Acts on order	Acts on order	Acts on order
4	Refuses order	Hesitation	Hesitation	Acts on order	Acts on order
3	Breaks	Refuses order	Hesitation	Hesitation	Acts on order
2	Breaks	Breaks	Refuses order	Hesitation	Hesitation

of the battle and by making the units' will-to-fight dynamic contingent on events in the game. Using such a rigid approach has the advantage of providing traceable results, should players dispute the outcomes, and it can be useful in generating data across games in a consistent manner to support comparative analysis.

Surprise

The defeat mechanisms discussion in Chapter Five touched on the idea of incorporating surprise into a wargame. Although players are often surprised in wargames, either by adversary players or by the game itself, explicitly designing surprise into wargames is an undertreated topic. Here, we discuss a theory of military surprise and role of surprise in wargames.

Theory of Surprise

Virtually all military theorists and practitioners rightly value surprise. Fuller included surprise as one of his nine principles of war, and most other lists of principles of war that we have reviewed have included surprise or at least subordinated it to another principle.[7] In terms of defeat mechanisms, surprise can both dislocate and preempt, two mechanisms that have their own logic and often convey advantage, allowing a force to subject an opponent to additional defeat mechanisms.[8] The preemption associated with surprise also includes gaining the initiative, where the initiative is another frequently lauded virtue in warfare.

Empirical study of surprise has revealed that, if surprise is achieved in battle, the probability of it leading to success is independent of force ratio and that the surprised defender usually surrenders, withdraws, or breaks—and does so more quickly than might otherwise be the case.[9] In addition to increasing the probability of success, surprise also reduces attacker casualties and increases the likelihood of inflicting shock.

In dissecting the impact of surprise at scales other than whole battles, surprise has different effects at different levels. At the tactical edge, surprised troops and their immediate leaders suffer stress, increased uncertainty, and information overload; this slows the speed of decisionmaking and often leads to poor decisions.[10] When commanders and staffs at higher echelons are surprised by enemy action but not, themselves, directly in contact with the enemy, there are different consequences: a tendency

---

[7] Fuller, 1926, p. 221; Leonhard, 1998, p. 147, for example, explicitly calls out surprise as important but subordinates it as a combination of two of his principles: *dislocation and confrontation and distribution and concentration.*

[8] Leonhard, 1998.

[9] Storr, 2009.

[10] Storr, 2009, p. 85.

toward command paralysis, a tendency to focus on details rather than the big picture, and an inclination toward micromanagement.[11]

Surprise has primarily mental and morale effects. To be surprised, understanding of the battlefield and enemy intentions must have been in error, and decisionmaking is thrown into disarray while old plans are discarded and new plans must be prepared and shared, likely with insufficient time (the mental sphere). Surprise is inherently dismaying, shaking confidence in commanders and in collective knowledge of the battlespace, dealing a blow to morale. There are no direct effects from surprise in the physical sphere, though surprise often accompanies a position of advantage from which disproportionate physical damage can be inflicted, which, in turn, compounds effects in the mental and morale spheres.

Surprise can be achieved in several ways. Fuller described three, corresponding to each sphere of war: surprise through superior direction (mental), superior determination (morale), or superior mobility (physical).[12] Storr highlights the unexpected as the central characteristic of surprise and notes that unexpected timing, direction, or means or methods of attack can result in surprise.[13] He adds that deception, intelligence, security, speed, and originality can be significant contributors in achieving surprise. Leonhard argues that surprise is easier to achieve than one might think, because military forces exist in a state of perpetual unreadiness and only climb briefly into a state of readiness when given forewarning and time.[14] Thus, the condition of surprise is part of the interplay between perpetual unreadiness and time, and surprise can be achieved, fundamentally, in only one of two ways: by delaying detection or by hastening contact.[15]

One closely related issue is deception; in practice, this is often how surprise is achieved. When employing deception, a commander develops a stratagem designed to mislead the opposing commander to either do something that would be disadvantageous or not do something that would be advantageous. In the words of Barton Whaley, "The purpose or goal of stratagem is to insure that the victim be surprised—that he does indeed choose a false or unfavorable alternative."[16] Generally, there are three logics behind stratagem:

1. Manipulate the perceived alternatives to eliminate choices that would benefit the adversary so that only suboptimal options remain.

---

[11] Storr, 2009, p. 85.

[12] Fuller, 1926, p. 273.

[13] Storr, 2009, p. 84.

[14] Leonhard, 1998.

[15] Leonhard, 1998, p. 188.

[16] Barton Whaley, *Stratagem: Deception and Surprise in War*, Cambridge, Mass.: Center for International Studies, Massachusetts Institute of Technology, 1969, p. 139.

2. Manipulate the perceived costs or benefits of various options and thus their attractiveness to the adversary.
3. Take actions that the adversary did not anticipate and thus could not respond to in a timely manner—relying on the element of surprise.

A range of stratagems can be used to achieve these effects:[17]

- camouflage and concealment, using natural or artificial materials to evade detection or hide from observation
- diversion, including demonstrations and feints that draw a target's attention away from an area or activity
- decoys or disguises, which use natural or artificial constructs to mimic an entity of significance, creating the perception of something that is not actually present
- sensory saturation, such as dazzling, with the goal of overloading the processing abilities of adversary equipment or personnel by drowning out a target signal with noise
- disinformation, in which doctored media are passed to the target
- exploiting a target's biases (either existing or introduced as part of the stratagem)
- disrupting decisionmaking by targeting either the adversary's leadership or C2 networks and interrupting the flow of situational awareness data or commands
- managing the timing or use of force to constrain the courses of action available to an adversary
- targeting the will to fight of leaders or troops, making them more willing to break from contact
- changing the behavior of partners and allies through diplomacy, changing adversary calculations.

Surprise in Wargames

Games are a natural way to explore surprise. Although modeling and simulation efforts can include decoys and deception, it is difficult for programmed agents to articulate the potential emotional aspects of surprise and how the experience of surprise affects decisionmaking.[18] In contrast, the human players in a game can actually experience surprise when playing against a thinking, reacting adversary force. What is more, players are able to articulate their expectations and reactions, and they can discuss intentions and perceptions explicitly in hotwash sessions. As a result, games provide a rich approach to studying surprise.

---

[17] Many of these examples are drawn from Scott Gerwehr and Russell W. Glenn, *The Art of Darkness: Deception and Urban Operations*, Santa Monica, Calif.: RAND Corporation, MR-1132-A, 2000.

[18] McConnell, 1988.

Surprise-themed wargames are predicated on unexpected and situational ignorance, either among players or among the units under their control. Fog of war is central to creating surprise, because players (or units) with perfect knowledge should never be surprised. These conditions can be challenging to simulate in wargames, however. The game would have to account for the effects of such factors as cyber operations, intelligence and counterintelligence, the additional complexity of the modern battlefield, or just the overwhelming amount of information available, conditions wargamers have called the "fog of data."

Commercial board games, often the inspiration for professional wargames, tend to have limited ways to deal with the fog of war because a boxed game can only do so much.[19] Simulating the fog of war can require double-blind games, in which neither side has perfect situational awareness or ground truth. However, double-blind games are more complex to plan, play, and adjudicate. For example, visualization and data representation in a double-blind game require separate views, rather than a common map or board that provides all players with the same information. The control team must also keep a separate record of ground truth for itself. All of this is challenging and time-consuming to execute manually and is often best handled through digital means. Adjudication is also more complicated because the control team needs to separately adjudicate results and determine what information from each result is available to each set of players. Finally, double-blind games can result in cautious decisionmaking and slow game play.[20]

To capture surprise effectively, games must have options for hidden information to affect situational awareness, player decisions, and adjudicated outcomes. For example, there must be some kind of credible way to seek and adjudicate delays in detection. This presumably requires a mechanism for tracking hidden movement and detecting and identifying enemy forces (using either a computer-based system for tracking and detecting unit locations or a system of multiple game boards). It also requires a mechanism to ensure verisimilitude in review, such that players who are told that their units are affected by surprise believe that delayed detection or hastened contact was legitimately achieved and not an artifact of unfairness in the adjudication system or gamesmanship on the part of the opposing players or—worse—the adjudicators.

In the case of manual games, the typical solution to this problem is the same as in 19th-century *Kriegsspiele*. Separate maps are maintained for each team, along with a "ground-truth" map for control. The control group (typically called the white cell) can then adjust the team maps to show only what can be reasonably observed through ISR and communicated with available C2 capabilities, allowing missing information,

---

[19] Discussion at the Fog of War Game Lab during the 2018 Connections Wargaming Conference, National Defense University, Fort McNair, Washington, D.C., July 19, 2018.

[20] Discussion at the Fog of War Game Lab during the 2018 Connections Wargaming Conference, National Defense University, Fort McNair, Washington, D.C., July 19, 2018.

incorrect information, and uncertainty in terms of the quality of information to effect player decisions. The downside of this approach is the considerable effort needed to vary situational awareness across multiple maps. Other approaches in commercial games involve the use of a single map alongside counters that limit the information available to an opposing player. Although this approach can be simpler to manage than the classic three-map approach, it limits what can be obscured—it is possible to hide information about the unit but not location—and can be logistically difficult in games with more than two sides. Computer interfaces can considerably reduce the burden of maintaining separate maps by managing rules for the fog of war, but this can require a substantial investment.

Capturing surprise in games also requires introducing appropriate consequences in the mental and morale spheres and, secondarily, in the physical sphere. Once a surprising force or action is detected, there should be appropriate delays at the appropriate organizational levels while the surprise is confirmed, adjusted to, and responded to. Appropriate dismay, discombobulation, and, perhaps, shock should be evaluated and imposed on surprised units, with attendant adjustments to likely casualties, increased chances of breaking and fleeing from surprise combat, and so on. These consequences will vary, depending on the stratagem selected by players. Table 6.4 suggests some potential mechanisms for adjudicating the effects of surprise in each of the three spheres.

Having described the important roles played by the mental and morale spheres (and potentially OIE) in defeat mechanisms, breakpoints, and surprise, we now turn to solutions that leverage these insights to help meet the requirements for bringing OIE into wargames.

Table 6.4
Incorporating Stratagem into Operational Games

Stratagem	Game Representation	Example Impact in Game Play	Likelihood of Impactful Use
Camouflage and concealment	Hiding real forces from the adversary Adjudicating successful deception	Hiding key assets Changing the perceived balance of forces in an area	Possible
Diversion	Normal force rules	Drawing off forces from one area to another	Likely
Decoys	Showing fake forces to adversary Adjudicating successful deception	Imitating key assets Changing the perceived balance of forces in an area	Possible
Disguises	Hiding real or showing fake forces, plans, or information to players Adjudicating successful deception	Imitating key assets Changing the perceived balance of forces in an area	Possible
Sensory saturation, such as dazzling	Blocking sensing capabilities of particular systems in particular areas	Preventing battle damage assessment over key bases	Possible
Disinformation	Showing fake forces, plans, or information to players Adjudicating successful deception	Drawing off forces from one area to another	Likely
Exploiting target's biases	Normal force rules	Drawing off forces from one area to another	Possible (needs longer play time)
Disrupting decisionmaking by disrupting leadership/C2	Limiting players' ability to control their own forces and learn ground truth	Delaying operational orders from reaching theater forces	Possible
Managing timing of use of force	Normal force rules	Preemption	Possible
Subverting or undermining will to fight	Rules for forces to break	Units failing to fight as expected	Low at the operational level
Changing behavior of partners or allies through diplomacy	Including green teams	Allies allowed to switch sides or enter or exit the conflict	Often treated as outside of the control of players

CHAPTER SEVEN

Additional Solutions to Address Gaps in Wargaming Operations in the Information Environment

The preceding chapters articulated a sequence of arguments. In Chapter Two, we argued that OIE are important, wargames are important, including OIE in wargames is important, and the current state of common practice for including the IE in wargames is deficient. In Chapter Three, we identified requirements for including the IE in wargames, beginning with six general categories of information in warfare that need to be addressed and concluding with Marine Corps–specific requirements drawn from stakeholder interviews and Marine Corps IRCs. We then argued in Chapter Four that the common failure of wargames to include OIE and the IE stems from implicit assumptions about the nature of conflict and mechanisms for defeat. Chapters Four and Five offered alternative assumptions, emphasizing Fuller's three spheres (mental, morale, and physical) as balanced contributors to combat outcomes and presenting a range of possible defeat mechanisms that include all three spheres and that, if employed in wargames, would accommodate the relevance of OIE and the IE. Chapter Six discussed breakpoints and surprise. Collectively, Chapters Four, Five, and Six offered practical advice for bringing these concepts into wargames. This chapter ties together these arguments and notes a few additional challenges to bringing OIE and the IE into wargames. We conclude by offering additional practical solutions for overcoming the various challenges presented in this report and meeting the requirements for representing OIE in wargames.

Challenges to Bringing the IE into Wargames

Given the importance of wargaming and the importance of OIE, why is the IE entirely absent from most wargames? Why are designers so inclined to represent only the physical and not the mental and morale sides of the equation?

Because games are generally built according to sponsor needs, one reason that mental and morale considerations have traditionally been marginalized is that there is not a demand signal for their need. Too often OIE are an afterthought, or just not a priority, for commanders.[1] Second, game designers often lack conceptual models of mental and morale phenomena on which to draw. Third, and perhaps related to the second, designers often depend on players' and expert adjudicators' mental models of the mental and morale dimensions of war, which makes the treatment of these issues subject to interpretation within the game and difficult to detect after the fact. Fourth, existing approaches to modeling mental and morale factors in conflict often remove player agency, which has consequences for game feasibility. We elaborate on these points in the discussion that follows.

As noted, designers lack good models of mental and morale considerations that they can incorporate into games. The wargaming literature often presents very general models that lay out very broad relationships. Applying such models to a specific context requires a great deal of work to identify what factors are important, how factors relate to one another, and how they shape the combat environment and results. Alternatively, historical case studies provide concrete details to help designers but are difficult to generalize to new contexts. What is missing is a set of principles that are general enough to be ported into games but specific enough to be credible—principles governing very broad general relationships as well as highly detailed but case-specific relationships.

This gap is particularly apparent when comparing how wargames represent the mental and morale spheres to how they capture the physical aspects of combat, in which there is greater alignment of key dynamics. Without an appropriate model of mental and morale considerations, the designer is left with several unattractive choices. One option is to model the resolution of all issues using a "lowest common denominator" model—in other words, to model causality in the physical world with only as much granularity as can be applied to the mental and morale aspects of play. Although such an approach may work for strategic games, which are typically played at high levels of abstraction, it will not lead to useful results in an operational or tactical game. The other option is to accept that the resolution of the models governing various domains and environments will not operate in the same way. Running such a multilevel game successfully is incredibly challenging: It is all too easy for the game to feel disjointed or for some aspects of game play to be marginalized, reducing the usefulness of results. Faced with these choices, it is easy to understand why a designer would prefer to leave mental and morale considerations outside the scope of game play.

In the absence of good models, designers often depend on players or adjudicators to depict the mental and moral aspects of war. In some ways, this is consistent with the role-playing element of games. For the particular echelon the player is representing,

---

[1] See Dennis M. Murphy, *Talking the Talk: Why Warfighters Don't Understand Information Operations*, Carlisle Barracks, Pa.: Center for Strategic Leadership, U.S. Army War College, Issue Paper 4-09, May 2009, and Christopher Paul, Colin P. Clarke, Bonnie L. Triezenberg, David Manheim, and Bradley Wilson, *Improving C2 and Situational Awareness for Operations in and Through the Information Environment*, Santa Monica, Calif.: RAND Corporation, RR-2489-OSD, 2018.

the player's decisions are the model of will and thus are informed by the players' own models of the mental and morale spheres. Similarly, adjudicators may bake in mental and morale considerations. For example, adjudicators in an attrition-based game will often state that they considered morale when constructing units' breakpoints, so what looks like a purely physical attrition calculation actually employs a richer perspective. Of course, this is unavoidably a poor abstraction. Even if mental and morale considerations are baked into a breakpoint calculation, the only way to achieve a break in the game is by inflicting casualties up to the breakpoint. A static baked-in model will not include a mechanism for changing the breakpoint by inflicting damage in the mental or morale sphere or through other effects.

Games, like all models, simplify and abstract some aspects of reality to allow a clearer understanding of key aspects of the system. In the case of a wargame, which seeks to represent days, weeks, or months of conflict or competition in hours of play, the representation of the environment, actors, and rules governing the effects of their interactions must be quite abstract to be playable. One result of this simplification is that the amount of information generated by the game will always be less than what is produced in the real world. As a result, the IE will always be somewhat distorted and simplified. These distortions do not mean that nothing can be learned about the IE or that the IE will have no effect; after all, the same point about abstractions applies to all other aspects of the game environment. However, when abstractions are made without considering their effect on the IE, it will be much more difficult to draw credible insights about OIE from the game.

In some areas, incorporating IE aspects into wargames will pose some special challenges. Representing partial situational awareness, deception, and the fog of war in games can require additional effort. Representing some very technical aspects of OIE, such as C2 networks, cyber capabilities, electronic warfare, and spectrum management, will likely require more-technical and -specialized tools. Visualization tools for OIE in wargames are another notable challenge, because such tools do not yet exist in actual OIE and are a real-world gap. Additionally, studying the differences between game audiences and adversaries and their real-world counterparts (to better understand likely information effects or lack of effects) requires specialized expertise and resources to establish it.

Yet, underlying these wargaming requirements is something more fundamental, a foundation to build upon that may not exist: good understanding and supportable theory about cause and effect in the IE. Attrition-based wargames incorporate physics-based conceptual models about cause and effect. However, for operations OIE, the correct causal relationships to build into wargames are not always as obvious or straight forward. As discussed earlier, building in unchallenged or incorrect assumptions risks creating negative learning and merely reinforcing existing biases. Game design, game mechanics, tool design, and other considerations all depend on a foundation of good understanding and a supportable theory of cause and effect.

Practical Crawl, Walk, and Run Solutions That Bring the IE into Wargames

Table 7.1 lists the six categories of information that are relevant to operations (and wargames), presented in Chapter Three, and describes what will be required to represent those categories in wargames. These solutions are presented according to a crawl-walk-run hierarchy, with the crawl level offering at least some way to consider the category in game design or play and the walk and run levels consisting of solutions to do so with greater sophistication and fidelity. For games with certain target levels of abstraction or objectives, the crawl- or walk-level solutions may be sufficient.

Table 7.1 should help game designers choose features to add that better represent the categories of information that are important to their sponsors (or to the purpose of a game). Table 7.2 connects various solutions to specific questions related to game design while drawing on requirements for representing OIE identified in Chapter Three. Many of these solutions overlap with those in Table 7.1; references to category solutions in Table 7.2 refer to Table 7.1.

Table 7.1
Crawl, Walk, and Run Solutions to Incorporate Six Categories of Information into Wargames

Information Category	Crawl Solution	Walk Solution	Run Solution
1. Situational awareness and situational understanding	Double-blind games with three maps (one "ground truth" held by the control cell and one map each for red and blue); rules for detecting, sensing, and updating; and manual adjudication	Digitized double-blind control for displaying what players should see	Sophisticated models that track not only locations but also, e.g., unit-sensing or -detecting radii, discrimination capabilities, and different levels of detection or identification
2. C2, including communication	SME-imposed delays of updates to maps, delays or degradation of messages, or no new orders to units outside the reach of communication systems	Systematic structure, rules, or tables dictating time required to see effects or receive messages, based on battlefield and IE conditions	Software supported by sophisticated models for, e.g., update delays, lags or disruptions in communications
	Physical separation of players at different locations or echelons, with intentional time delays in communication between them, as appropriate	Software that imposes rule- or table-based delays in updates to information	

Table 7.1—Continued

Information Category	Crawl Solution	Walk Solution	Run Solution
3. C2W and other factors that degrade situational awareness or C2	Crawl solutions for categories 1 and 2, plus additional rules or SME adjudication of C2W and other IRC actions to increase fog of war or delays or to affect perceptions	Walk solutions for categories 1 and 2, plus additional software rules for C2W actions, their adjudication, and their effects	Run solutions for categories 1 and 2, plus additional complex software that adjudicates specific networks and their connections and presents both the direct impact of various C2W actions and their cumulative or interactive effects
	Log of physical environment and IE characteristics so that players who are deceived or outflanked can see (and accept) how game play actually unfolded	Detailed software-based log of physical environment and IE characteristics so that players who are deceived or outflanked can see how game play actually unfolded	
4. Information that causes subordinates to behave in ways contrary to the commander's orders or preferences	Manual rules, guidelines, or tables for the behavior of units outside the reach of command, leadership or command tests (and situational modifiers), and the quality and characteristics of subordinates	Software-based tracking of circumstances leading to crawl-level situational modifiers and digital tracking of leader characteristics or states and digital tests for compliance	Sophisticated simulation, with a full simulation needed for multiagent models and complex decision rules
	SME injection or adjudication when circumstances cause subordinates to behave differently		SME input conveying deep understanding of human dynamics, the cognitive and cultural biases of relevant decisionmakers, and other behavioral factors
5. Information for effect	Crawl solution for category 4, plus rules or adjudication for efforts to affect or exploit circumstances	Walk solution for category 4, plus digital tracking or incorporation of effects based on relevant circumstances	Run solutions for categories 1, 2, and 4, when unit actions of reflect, e.g., characteristics and proclivities, awareness and perception of the situation, and last orders received (and how long ago they were received)
6. Factors that affect relevant actors other than adversaries	SMEs in control cell loosely track relevant actors, their changing sentiments, and their actions in response to OIE and other actions or events	Mechanism to track all relevant actors and their states and conditions; systematic structure or rules for how game actions or events can affect those states and conditions; systematic structure or rules for the resulting consequences of relevant actors' actions in the game	Sophisticated software-based modeling, including advanced social science, narrative, human dynamics, and cognitive and cultural biases

Table 7.2
So You Want a Game That . . . ?

Challenge	Solutions
Includes IE-related scenarios	• Provide a "road to conflict" and other background materials that include the IE and describe it as relevant. • Include IE-related objectives in game objectives or victory conditions. • Include IE-related conditions and stimuli that encourage players to think about the IE. • Script or encourage the red team to execute OIE alongside kinetic operations. • Explicitly include IE-related actions in lists of possible game actions.
Captures OIE with greater depth and realism	• Include the possibility that actions in the IE might fail, and outline the consequences for the failure of OIE (or the failure to conduct OIE). • For IRCs, consider the probability of success and possible reasons for failure or partial success, and include a random component similar to whatever adjudication mechanism determines the success or failure of other game actions. • Incorporate solutions to other OIE-related challenges as needed.
Captures the effects of OIE on noncombatant populations	• Incorporate recommended solutions for category 6 information. • Acknowledge the presence of noncombatants and their role in the game, and incorporate these points into SME adjudication. • Adopt some kind of model of states (sentiments and actions) of various groups, rules for state changes, and rules for actions in different media (e.g., social media, print, TV, radio), and enlist SMEs with expertise in civilian behaviors.
Captures the effects of noncombatant populations on the IE and operating environment	• Proceed as in the row immediately above with a greater emphasis on actors' in-game actions.
Captures the impact of others (e.g., coalition partners, other green forces) in the IE	• Proceed as in the row immediately above but representing a different set of relevant actors. • Add a player or adjudicator for each relevant partner, entity, or group
Provides situational awareness of the IE	• Incorporate recommended solutions for category 1 information. • Adopt visualization tools that mirror output from real-world tools. (At the walk level, such tools would be hand-jammed by SMEs; at the run level, computationally intensive simulation would replace human intervention.) • Include the IE in intelligence summaries, updates, adjudication, and other materials provided to players.
Provides feedback to players related to the IE and OIE	• Proceed as in the row immediately above. • Allow structured adjudication of OIE, and enlist SMEs with a greater depth of expertise in human dynamics. • Create an after-action report highlighting relevant information on events and trends in the IE versus what was observable to players. • Provide feedback in the form of realistic OIE assessments.

Table 7.2—Continued

Challenge	Solutions
Allows assessment of OIE	• Provide changes in ground truth filtered through an assessment process and represented in assessment metrics; no assessment plan means limited or no feedback. • Create an after-action report that shares what actually happened in response to OIE versus what was observed through assessment.
Captures IE effects on combat	• Use a three-sphere damage-adjudication approach, and allow a wider range of defeat mechanisms.
Includes C2W	• Incorporate recommended solutions for category 3 information.
Includes SIGMAN	• Incorporate recommended solutions for category 3 and 4 information. • Communicate the possible consequences of successful, partially successful, and failed SIGMAN. • Allow hidden movement. • Include the impact of SIGMAN in red-team indications and warnings and situational awareness.
Includes deception or surprise	• Incorporate recommended solutions for category 1, 3, and 4 information that allow the game to represent the fog of war. • Enforce mandatory behavior on units to represent the effects of surprise on lower echelons.
Includes cyber and other technical capabilities	• Match timelines and level of abstraction with desired technical play. • Ensure that the adjudication of effects that cyber operations might plausibly deliver.
Includes electronic warfare	• Incorporate recommended solutions for category 2 and 3 information. • Build electronic warfare into physical simulation models, as appropriate.

CHAPTER EIGHT

Conclusions and Recommendations

We conclude this report with a summary of findings and conclusions, followed by recommendations for specific groups of stakeholders: those who sponsor wargames, those who design wargames, and those who procure and provide wargaming capabilities.

Conclusions

Wargaming Has the Potential to Inform the Marines Corps as It Works to Develop Better Approaches to Operating in the IE

The IE and OIE are increasing points of emphasis within the DoD and the Marine Corps. Information has joined the list of joint functions, and the Marine Corps is in the process of adding it to its own list of warfighting functions. New concepts related to OIE have emerged at both the joint and service levels, and the Marine Corps is so committed to OIE that the Commandant established the position of Deputy Commandant for Information and created MIGs at each of the MEFs.

Wargaming Can Be Incredibly Valuable, but Its Value Is Diminished if It Does Not Include the IE

This is obvious for IE-related issues (you cannot wargame the IE unless you include the IE), but is also important for gaming operations that do not emphasize the IE: If the IE is absent, games risk being unrealistic, potentially leading to incorrect conclusions and wrong lessons and observations. The IE and OIE are not Marine Corps–specific or IRC-specific considerations; they are instead fundamental aspects of the operating environment that anyone engaged in wargaming should consider.

Unfortunately, the IE, OIE, and IRCs are all frequently underrepresented in existing wargames. Although OIE and the IE have garnered increased attention in joint and Marine Corps concepts and doctrine, equivalent attention has not been paid to representing OIE in wargames, limiting the opportunity for this tool to contribute to new concepts, foster innovation, and instruct the force of the future. Common wargaming practices, and existing literature and tools currently offer minimal support for OIE. When wargames do consider information, they most often focus on situational awareness, fog of war, and C2. Attrition-based defeat mechanisms predominate in the current generation of defense wargames, and these games do not typically allow effects in and through the IE to influence combat results or other outcomes.

Although There Is a Substantial Need to Incorporate OIE in Wargames, *There Are Barriers to Doing So*

There is an absence of well-developed and generally accepted abstract models for OIE and effects in and through the IE. Physically focused abstract models for combat are simply more mature. Such models will need to be developed before IE can be included in wargames at the run level in our crawl-walk-run progression. There is reason for optimism, however: Wargaming's physical combat models survived and evolved despite the deeply flawed Lanchester Equations that informed early combat modeling.[1]

Recall that at least six types of information are relevant to military operations and worth considering when designing wargames:

1. situational awareness and situational understanding (including battle damage assessment)
2. C2, including communications
3. C2W and other factors that degrade situational awareness or C2
4. information or aspects of the IE that can cause subordinates to behave in ways contrary to the commander's orders or preferences
5. efforts to protect against contrary subordinate behavior or to encourage such behaviors among adversary forces (information for effect)
6. factors in and through the IE that affect the perceptions or behaviors of relevant actors other than adversaries.

Military Theory Provides Some Possible Solutions

Military thinking from J. F. C. Fuller to contemporary Marine Corps doctrine acknowledges three spheres in which "defeat" occurs: the mental, morale, and physical. Most wargames (and other forms of military analysis) focus only on the physical, but the first two are critical not only to depict OIE but also to accurately capture the nature of conflict. Considering these three spheres leads quite naturally to defeat mechanisms that move beyond attrition. We highlighted ten additional defeat mechanisms from military theory, all of which could be included in games. Furthermore, the same sort of three-sphere adjudication mechanics that would allow OIE to be meaningfully represented in wargames would also capture the impact of physical combat actions even better than current attrition-based models.

---

[1] These equations, developed in the 1950s, represented the strength of each side in a wargame as a constant effect of time and relative physical strength (so, they are purely physical combat adjudication models). See the appendix in Paul K. Davis, *Aggregation, Disaggregation, and the 3:1 Rules in Ground Combat*, Santa Monica, Calif.: RAND Corporation, MR-638-AF/A/OSD, 1995.

To better depict OIE, games need to do the following:

- represent some or all of the six different types of information relevant to military operations
- provide information to players to inform their actions
- allow players to take actions in the IE
- have a system that allows player actions in the IE to meaningfully effect other players and the physical environment
- include tools to provide feedback about these actions to players
- include player decisionmaking as a focus of data collection and analysis
- represent and consider all three spheres of war
- include a wider range of defeat mechanisms.

Recommendations

General Recommendations

We begin with two general recommendations for any defense wargaming stakeholders. Everyone involved in wargaming should acknowledge the role of information in operations and seek to better represent relevant aspects of the IE in games. Furthermore, stakeholders should recognize the potential role for OIE across the spectrum of conflict.

The effects of OIE do not remain in the IE; such operations have consequences in and across the spatial domains. OIE are not just useful in affecting noncombatants; these operations also have a role in combat against peer or near-peer adversaries. And OIE are not just relevant in phase 0; these tactics are applicable in all phases and at all intensities of conflict. Game accordingly.

Recommendations for Those Who Sponsor Wargames

Those who sponsor wargames set the objectives and requirements for games, communicating these needs to those who design the games and those who procure the tools and other support for gaming. If wargame sponsors demand more and better representation of the IE and OIE, other stakeholders will try to find ways to better meet those demands.

The following recommendations can help guide sponsors as they plan for games, evaluate designs, and oversee play and follow-on analysis:

- Explicitly emphasize the role of the IE and its relevance to the game's purpose.
- Move from critiquing the absence of OIE to demanding their inclusion in games. Require that games move away from attrition-centric models to models that more fully consider the mental and morale spheres of conflict in ways that allow players

to make meaningful decisions and that project plausible outcomes in meaningful ways.

- Explicitly require game designers to include OIE and make that requirement clear during scoping conversations.
- Explicitly require that OIE remain central to game play, and ensure that there is a procedure for documenting these operations and their effects in post-game analysis to enable evaluation and learning by players, other sponsors, and wargame designers.
- If the relevance of the IE is not evident at each stage of a game's design and execution, engage with designers to understand how OIE are being included.
- Be prepared to support additional recruitment efforts to bring in nontraditional players who can enhance the game's ability to adequately capture the IE. To further improve the fidelity of game play, consider sponsoring the development of OIE-related educational materials for players.

Although the wargaming field is building new approaches to representing OIE, it will be important to document game designs that represent these operations so that lessons and techniques can be shared across organizations. Whether wargames include OIE should be cited in these games' Defense Wargaming Alignment Group repository entries to enable other sponsors and designers to learn from ongoing efforts. The categories of MAGTF IE operations and types of information may be a helpful start to developing such a labeling scheme.

Recommendations for Those Who Design Wargames

Those who design wargames in response to sponsor requirements will ultimately be responsible for bringing the IE into their games and determining which aspects of the IE and OIE to represent, at what level of abstraction, with what possible in game actions, and with what possible impact.

The following recommendations will help designers consider the IE from the earliest stages of planning:

- Identify which of the six types of information are important to the overall purpose or objectives of the game and find ways to incorporate these interactions.[2]
- When developing new systems, start from the notion that the mental, morale, and physical aspects of a conflict need to be represented. Because the tools for representing physical aspects are more established, holistically representing the conflict will require devoting substantial time to the development of mental and morale approaches.

---

[2] See Table 7.1 in Chapter Seven for potential solutions.

- Remember that representing more spheres of conflict will add more elements to the game and thus more complexity. Think hard about how to balance complexity and abstraction to most usefully characterize the key challenges within the game.
- Remember that IE expertise may not reside among the "usual suspects" in the wargaming community. Be prepared to bring in new SMEs to help evaluate, test, and refine new game mechanics and underlying models
- Remember that members of all cells should have players who are familiar with OIE. Recruit accordingly, and be prepared to recruit from new communities or to build educational content into the game.
- Ensure that post-game analysis considers the aspects of the game's design or play that affected player decisionmaking, particularly with respect to information, and determine why players made the decisions they made.

Recommendations for Those Who Procure and Provide Wargaming Capabilities

Wargame design and execution requires various types of support, including the procurement of game materials, technologies, and expertise. If a sponsor states a requirement and a designer conceives a way to meet it, it may still not be possible to implement with sufficient fidelity or expertise if the supporting establishment has not invested in the right tools or expertise. When we describe those who procure and provide wargaming capabilities, we mean to cover the full range of possible capabilities—from the physical spaces for game play, modeled communication and network capabilities, the software tools, game board or maps, and scenario details and background materials to the design of opposing forces and supporting cells and the selection of SMEs and expert adjudicators. Our recommendations focus primarily on systems and software and necessary expertise, however.

The following recommendations apply to the procurement of software and tools to support game development or play:

- Select tools that can represent the mental and morale spheres and a range of defeat mechanisms, information, and conditions that can affect game play (which may be unknown to players), along with robust models of human dynamics, psychological factors, and information flows.
- Ensure that systems are transparent to allow designers to validate how OIE are incorporated and their effect on game play.
- Check that the weights of specific types of information, operational actions, and noncombatant behaviors can be updated and modified as understanding of OIE matures, and consider how these lessons can be applied to new regions and conflicts.
- Select systems that can represent the types of information needed by involving game designers and sponsors in the evaluation process.

The following recommendations are intended to guide the recruitment of personnel to support game design, testing, and execution or the involvement of SMEs in specific aspects of these tasks:

- Specify requirements for expertise in such areas as qualitative research and social science, evolving OIE practice, and regional populations when writing statements of work, task orders, and position descriptions.
- Representing OIE requires knowledge of relevant operational art, adversary doctrine and practices, and local context. It may not be practical to maintain staff with expertise in these areas at all times, but efforts should be made to recruit supplemental experts to support specific game requirements. Capability developers should form a "stable" of relevant IE and OIE expertise by requesting SME rosters from other U.S. military and government organizations and by developing relationships with outside experts.
- Because the concepts and practices associated with OIE are evolving, OIE expertise is not static. Experts should refresh their knowledge as needed to stay up to date with new practices and evolving contexts.

Directions for Further Research

Given the complex challenge of wargaming OIE, along with the substantial gaps and shortcomings common in current defense wargaming practice, there are several areas that warrant further research and development.

One major need is for a systematically compiled body of knowledge about effects in the IE and what is known about effects in the IE that can be generalized to models for game design and the choice of game mechanics. In the absence of a robust collection of knowledge based on lessons learned, after-action reports, historical case studies, and other assessments of past incidents and conflicts that have included OIE, it is difficult for wargame designers to understand exactly what defeat mechanisms and effects to build into wargames. How did information affect major combat operations, and aspects of the IE are pivotal for current wargames against near-peer adversaries? What are the potential real-world effects of certain IRC activities? How did various contexts and other factors contribute to success or failure? How have adversaries and audiences changed or adapted over time? Given that communication methods and tactics are constantly changing, do the lessons learned five or ten years ago still hold? Have actors learned from past U.S. attempts to influence or deceive, and how have they adapted or attempted to counter U.S. actions?

Having this type of research already on hand when beginning to design a wargame would improve how OIE and the IE are represented, since many wargame designers and their staffs are not OIE experts and have only limited time to organize their events.

Having a body of knowledge about past IE effects to review and critique would also help mitigate the tendency to mirror past wargame designs or to rely on biased assumptions and conventional wisdom when it comes to modeling the IE. Again, this is critical to avoiding negative learning in games. Such a body of work would also help clarify when it is appropriate to apply lessons learned in one context to another. Because the combat experience in the Marine Corps is rooted predominantly in counterinsurgency operations in Iraq and Afghanistan, lessons learned from other types of conflicts would be helpful when Marines are called upon on participate in other types of wargames—those focused on conflict dynamics with which they have no direct experience and for which there are potentially few experts to serve as wargame adjudicators.

Another area for further work is creating wargame scenarios with improved IE backstories and additional hooks for information and its potential effects. Sharing a variety of potential wargame scenarios that include OIE in a range of military operations and at different levels (tactical, operational, strategic) may be a practical first step toward bringing information considerations to the forefront in games that would otherwise downplay the role of information. Another benefit to information-centered scenarios is that they can preempt shortfalls in general wargaming practice. Put another way, if static scenarios and scripted adversary players continue to be a standard approach to wargaming in DoD, an immediate step for bringing OIE and the IE to the attention of wargame sponsors, designers, capability procurers, and players is to make them integral to the scenario and to adversary capabilities or intentions. One approach is to conduct a wargame with the explicit purpose of creating such a scenario—a "game before the game" that would allow key actors to move in the IE. Doing so would create a deeper information-focused backstory and provide greater context to discuss and adjudicate information effects.

Another key area for future research is how the information presented in a game affects player decisionmaking. In games in which information is central, understanding how players react to that information is also central. In this report, we argued that the mental factors in war are often overlooked in typical wargames. Better understanding and analysis of players' decisionmaking, perceptions, mental models, misperceptions, and reasoning can substantively enhance understanding of how information is used or not used in games and how players' perceptions or actions change as a result of the information presented in a game.

This area of wargame analysis would benefit from greater specialization as well. Defense wargame analysts tend to represent technical disciplines rather than cognitive sciences, psychology, communication, narrative, or other nontechnical areas of expertise. Systematic observation and research on decisionmaking in wargames—including in the context of surprise maneuvers, partial situational awareness, and active deception—are necessary to better examine how information is incorporated into wargames.

Related to understanding how players respond to information is understanding how responses to information may differ between the world of a wargame and the real world. As anyone who has participated in wargames has realized, when other U.S. personnel stand in for the adversary, there is the perpetual question of how accurately they can imitate the decisionmaking of real-world adversaries. This is a challenging area because most wargame scenarios never actually play out in reality, leaving analysts with few non-hypothetical cases to compare actions and outcomes in a wargame versus in reality. Nonetheless, it remains important to assess how wargame stand-ins for adversaries and other actors may differ from their real-world counterparts in an effort to addresses potential biases and mitigate negative learning from games. Work in one area that we have already covered—better understanding of actual OIE—should help as well. However, contemporary actors will differ from past ones, and future actors will be even more difficult to assess objectively.

Another important area for future research concerns recommendations for wargaming tools. As the Marine Corps begins building its new wargaming center in Quantico, Virginia, it intends to invest a great deal of resources in modeling and simulation and advanced wargaming tools. Additional research on how these tools can or should incorporate OIE and the IE before procurement decisions are finalized may help inform and enhance Marine Corps OIE wargaming in the long run.

APPENDIX A

Wargaming Glossary

This appendix offers a brief primer on wargaming terminology and is meant to provide an overview for those unfamiliar with wargaming. Wargaming as a field is somewhat immature, so the terminology is not fully standardized, and usage varies among practitioners. In seeking common definitions, we drew heavily on excellent compilation *A Compendium of Wargaming Terms* by William L. Simpson, Jr., which, itself, draws on multiple other sources (also cited here), but we have also amended and refined these definitions.[1] Those interested in alternative perspectives would be well served by referencing Simpson's guide.

Wargaming and Related Analytical Tools

Exercise: "A military maneuver or simulated wartime operation involving planning, preparation, and execution. It is carried out for the purpose of training and evaluation."[2] Subtypes include the following:

- **Command post exercise:** "An exercise in which the forces are simulated, involving the commander, the staff, and communications within and between headquarters."[3]
- **Field exercise:** "An exercise conducted in the field under simulated war conditions in which troops and armament of one side are actually present, while those of the other side may be imaginary or in outline [i.e., abstracted]."[4]

[1] William L. Simpson, Jr., *A Compendium of Wargaming Terms*, last updated September 20, 2017.

[2] W. Simpson, 2017, quoting an earlier edition of JP 1-02, *Department of Defense Dictionary of Military and Associated Terms*, Washington D.C.: U.S. Joint Chiefs of Staff, current as of July 2019; the term had been removed from the latest edition of JP 1-02 at the time of this writing.

[3] JP 1-02, 2019; the language is quoted in W. Simpson, 2017.

[4] JP 1-02, 2019; the language is quoted in W. Simpson, 2017.

Model: "A physical, mathematical, or otherwise logical representation of a system, entity, phenomenon, or process."[5]

Simulation: "A method for implementing a model over time."[6]

Tabletop exercise: Used as a synonym for *wargame*.[7]

Wargame: "Adversarial by nature, wargaming is a representation of military activities, using rules, data, and procedures, not involving actual military forces, and in which the flow of events is affected by, and in turn affects, decisions made during the course of those events by players acting for all actors , factions, factors and frictions relevant to those military activities."[8]

Workshop: "Workshops involve subject matter experts (SMEs) gathered to discuss a problem. Workshops have a narrow, discrete focus, and often serve as an input to follow-on [Wargaming Division] events."[9]

Types of Games

Analytical wargame: "Competitive, contextualized decision-making within pre-defined constraints for the purpose of gaining insight into complex, adaptive, interactive, and cognitive systems."[10]

Board game: A game played on a board, such as chess or Monopoly.

BOGSAT (a bunch of guys and gals sitting around a table): A somewhat disparaging term used to describe seminar-style gaming, suggesting that nothing worthwhile can be accomplished.

---

5 Defense Modeling and Simulation Coordination Office, U.S. Department of Defense, *DoD Modeling and Simulation Glossary*, September 24, 2018, quoted in W. Simpson, 2017.

6 Defense Modeling and Simulation Coordination Office, 2018, quoted in W. Simpson, 2017.

7 Matthew Caffrey, lecture delivered as part of Air Force Materiel Command Wargame Course, c. 2016, quoted in W. Simpson, 2017.

8 Paul W. Vebber, Assistant Director, Wargaming and Future Warfare Research, Naval Sea Systems Command Warfare Center, Undersea Warfare Directorate, "Wargaming Compendium," email to the authors, July 8, 2015.

9 Burns, undated, quoted in W. Simpson, 2017.

10 Jon Compton, wargame designer, "Toward an Epistemology of Wargaming—A Drunkard's Walk," briefing, Military Operational Research Society, Wargaming Community of Practice Brown Bag Lecture Series, March 19, 2015, quoted in W. Simpson, 2017.

Closed wargame: A wargame in which "the 'fog of war' limits what the players know of thc total situation depending on their 'cell.' In closed wargames Wargame Control determines what the players should logically see and know about the real situation using the 'Three Map' method. Poker is an example of a closed wargame."[11] "A game in which players receive the amounts and kinds of information and intelligence of friendly and enemy forces that they would normally receive in a similar real-world situation. Most war games are closed games."[12]

Commercial game: A game sold for recreational play. The commercial wargaming market is often segmented by game medium—for example, video, console, computer, mobile, board, card, miniature, and role-playing games. Commercial games are an important source of inspiration and new techniques and can sometimes be used directly to support research and educational objectives.

Educational game: "A game conducted to provide military commanders or executives with decision-making experience, and to familiarize them with the operations and problems involved."[13]

Kriegsspiel: German for *wargame*. Nineteenth-century Prussian *Kriegsspiele* that assigned players to teams and used maps and playing pieces to represent battlefield conditions and unit types and locations are the precursor to today's wargames.

Manual games: "A game in which the forces are represented by models, pins, pieces, or symbols, and the participants move them about by hand on a chart, map, board, or terrain model which represents the area of operations."[14]

Matrix game: A game in which players are instrumental in determining the outcomes of moves by using a matrix adjudication process.[15]

Miniatures wargame: "Type of wargame played on a [three-dimensional] model of terrain with [three-dimensional] representations of the engaged forces."[16]

---

[11] U.S. Marine Corps Warfighting Laboratory, Wargaming Division, *Standard Operating Procedures (SOP)*, August 2005, quoted in W. Simpson 2017. The three-map method consists of three views of the battlefield: a ground-truth view held by the white (or control) cell and one map each for the red and blue sides.

[12] McHugh, 2013, quoted in W. Simpson 2017.

[13] McHugh, 2013, quoted in W. Simpson 2017.

[14] McHugh, 2013, quoted in W. Simpson, 2017.

[15] See Curry and Price, 2014.

[16] Caffrey, undated, quoted in W. Simpson 2017.

Open game: "A game in which all players receive or have access to all information and intelligence of the actions of all friendly and enemy forces. Usually played in one room and on a single map or chart."[17]

Seminar wargame: A game format in which "opposing players discuss the sequence of moves and countermoves likely to be made in a given situation and agree on interactions likely to occur. The control team adjudicates the results of those interactions and reports back to the players. The process is repeated for each of the 'moves' in the game."[18]

Title 10 wargames: "Any of a series of wargames conducted by each of the U.S. armed services addressing its duties under Title 10 of the U.S. Code to organize, train, and equip its forces."[19]

Elements of a Game and Design Choices

Adjudication: Various methods may be used to adjudicate, but, in general, adjudication is "the procedure to impartially resolve the outcome of interactions between sides in a game."[20] The term assessment is often used interchangeably with *adjudication*.[21] There are three kinds of adjudication:

- **Free adjudication:** "The results of interactions are determined by the [adjudicators] in accordance with their professional judgment and experience."[22]
- **Rigid adjudication:** "The results of interactions are determined according to predetermined rules, data, and procedures."[23]

---

17 McHugh, 2013, quoted in W. Simpson 2017.

18 Marine Corps Order 3070, *The Marine Corps Operations Security (Opsec) Program*, draft, June 1991, quoted in W. Simpson 2017.

19 Matthew B. Caffrey, Jr., *On Wargaming: How Wargames Have Shaped History and How They May Shape the Future*, Newport, R.I.: U.S. Naval War College, 2019, p. xxv.

20 Burns, undated, p. 51.

21 Simpson explains that *adjudication* is used as an alternative to *battle damage assessment*, which, "in wargaming has traditionally focused on the purely kinetic or physical damage aspects of warfare." He adds that battle damage assessment typically "ignores the non-kinetic factors dealing with knowledge, deception, situational awareness, information warfare, morale, public opinion, etc." (W. Simpson, 2017).

22 McHugh, 2013, quoted in W. Simpson 2017. McHugh refers to adjudicators as "umpires" and adjudication as "umpiring," common terminology at the time he was writing in the 1960s.

23 W. Simpson 2017, paraphrasing McHugh, 2013.

- **Semi-free adjudication:** "Interactions are evaluated by the rigid method, but the outcomes can be modified or overruled by the lead adjudicator."[24]

Cell: "A group of wargame participants organized to accomplish an assigned purpose."[25]

Combat results table: "A Probability Table that shows the possible results of all [actions] allowed within a particular game. . . . [A] die or other random-number generator is used to determine the actual result," to better represent the large number of potential influencers on outcome outside of player action. "These tables are usually calculated based on what information is available on actual historical losses" or, we would add, other outcomes or relevant theory.[26]

Counters (also called *tokens* or *units*): In recreational wargaming, forces are represented by counters. In board games, the convention is to use half-inch square pieces, appropriately colored and labeled, to represent anything from a single soldier to complete armies. Unit-level counters convey several pieces of information, including the unit type, its name, and its posture (offensive, defensive, or another movement indicator).

Fog of war: "The uncertainty of friendly, adversary, and environmental conditions experienced by leaders in armed conflict."[27]

Game board: Typically, a two-dimensional map-style representation of the terrain on which the game takes place. An overlay of squares or hexagons represents subdivisions of a potential area of operations.[28] In other cases, the game board can represent nonphysical terrain, such as a bureaucratic process or a social, informational, or technological network.

Ground truth: "The actual [position and] status of units played in a game. In an open game all players are given the ground truth. In a closed game the Control Cell maintains ground truth and communicates a limited and possibly distorted view of ground truth to the different sides playing in the game."[29]

---

[24] W. Simpson 2017, paraphrasing McHugh, 2013.

[25] Burns, undated, quoted in W. Simpson 2017.

[26] Dunnigan, 2000, quoted in W. Simpson 2017.

[27] Caffrey, undated, quoted in W. Simpson, 2017.

[28] See McHugh, 2013.

[29] U.S. Marine Corps, 2005, quoted in W. Simpson, 2017.

Hex grid: "The hexagonal pattern laid over a map to facilitate unit movement, measuring distance, and providing location of units."[30]

Hotwash: An important after-game activity involving either informal information discussion about the game's outcomes or formal analysis of the game's structure and play.[31]

Injects: Scripted events used as "supplements [to] the exercise scenario [which include] event synopses; expected participant responses; capabilities, tasks, and objectives to be addressed; and responsible personnel. It includes specific scenario events (or injects) that prompt players to implement the plans, policies, and procedures that require testing during the exercise, as identified in the capabilities-based planning process. It also records the methods that will be used to provide the injects. (i.e., phone call, facsimile, radio call, e-mail)."[32] The list of injects planned for a game is the master scenario events list.

Matrix adjudication: A specific approach to adjudicating seminar-style games in which both sides present arguments for the likely outcome of an action and the adjudicator determines the outcome that will actually occur in the game.[33]

Move mechanics or rules: "The sequence of actions or steps taken by the players or sides to carry out their moves and how they interact with each other during the sequence."[34]

Number of sides: "The number of sides in a game is determined by the nature of the conflict and the nature of the opposition being gamed and the number of independent entities who can make decisions and take independent action that influence the direction of the game."[35]

Order of battle: "The make-up of a player's forces in terms of types of units and the number of each type available."[36]

---

[30] W. Simpson, 2017, paraphrasing Dunnigan, 2000. The approach originated at RAND as a means of creating consistent movement in all directions without the distortions caused by diagonal movement on a square grid. See John F. Nash and R. M. Thrall, *Some War Games*, Santa Monica, Calif.: RAND Corporation, D-1379, 1952.

[31] See Perla, 2011, pp. 23–24.

[32] Defense Modeling and Simulation Coordination Office, 2018, quoted in W. Simpson, 2017.

[33] See Curry and Price, 2014.

[34] Joey Linehan, "Expeditionary Advanced Base Operations (EABO) Wargame 2015 Design Brief," planning conference briefing, Quantico, Va., September 10, 2015, quoted in W. Simpson, 2017.

[35] W. Simpson, 2017.

[36] Dunnigan, 2000, quoted in W. Simpson, 2017.

Role-playing: "Role-playing is whether or not the players are restricted to perform[ing] the duties of assigned game billets or . . . free to participate according to their experience and knowledge."[37] "Training wargames and games examining the functioning of staffs, [headquarters], or teams usually require strict role playing or role play."[38]

Scenario: The world in which the game is played, including the conflict situation that players must resolve. "A narrative description of the setting in which the strategic, military, political, economic and social environment is established and the physical geography is set forth. It may be hypothetical, real or a combination of both."[39] "Scenarios include information such as threat and friendly politico-military contexts and backgrounds, assumptions, constraints, limitations, strategic objectives, and other planning considerations. A scenario is intended to represent a plausible challenge and may not reflect the most likely events."[40] The modeling and simulation community defines *scenario* more broadly to include the identification of key actors and their capabilities, behaviors, and relationships.[41]

Turn-based wargame: "A turn based wargame is one in which Blue and Red can only affect the progress of the wargame at [discrete] times on a regular schedule. Ideally a wargame should allow player turns at the same interval as real world decision cycles."[42] Game turns can be managed in a number of ways, including alternating (one player takes a turn, then another), simultaneous (all players give orders to their units and actions unfold and are adjudicated simultaneously), or integrated (players alternate moving units as part of a single broader turn, subject to possible constraints, such as that they must activate a particular unit or activate units in a random order).

Vignette: "A brief descriptive sketch of a situation. A mini-scenario confined to a very specific and limited situation. A vignette can be a subset of a larger scenario."[43]

---

[37] U.S. Marine Corps, 2005, quoted in W. Simpson, 2017.

[38] W. Simpson, 2017.

[39] Marine Corps Order 3070, 1991, quoted in W. Simpson, 2017.

[40] U.S. Department of Defense Directive 8260.05, *Support for Strategic Analysis (SSA),* Washington, D.C., July 7, 2011, quoted in W. Simpson, 2017.

[41] See Defense Modeling and Simulation Coordination Office, 2018.

[42] Caffrey, undated, quoted in W. Simpson, 2017.

[43] U.S. Marine Corps 2005, quoted in W. Simpson, 2017.

Roles in a Game

Colors or cell identification: "Opposing sides and their allies in a game are normally assigned specific colors to reduce confusion. Red and Blue are the two basic opposing sides in a two-sided game," with blue generally representing the United States and its allies and red representing adversaries, particularly peer competitors, such as Russia and China.[44] Green is usually used to represent civilians, the environment, neutral third parties, an ally of the blue side, or the host nation. White can signify a variety of roles but most often indicates the game's facilitator or adjudicator, as discussed next.[45] Depending on the number of sides in a game, additional colors may be assigned.

Control cell or white cell: "The person or group of persons designated to monitor and direct game execution to meet the game objectives[,] also called game control."[46] Although this is the most common definition, *white cell* can also refer to key entities that are unable to participate but whose input is critical to game play or assessment and analysis.

Controller, umpire, or white cell member: "A member of the control group who performs one or more of the following duties: monitors player actions, evaluates interactions, [and] provides intelligence to players."[47]

Facilitator: In addition to guiding discussion among players in a cell, the facilitator "ensures that the cells accomplish their game objectives [and] provides [links] between the player cells and the White Cell, relaying guidance and questions, and ensuring that the cells stay on track." The facilitator also "provides insights and assessment" to the game's analysts.[48]

Higher authority: "Designated Controllers, who act as commands and organizations outside of game play and . . . make decisions and provide policy guidance to the players as required to maintain game play and keep the game focused on the objectives." The

---

44 W. Simpson, 2017. Simpson notes that this red-blue color-coding dates to 19th-century Prussian wargaming, adding, "When Russia and China adopted modern wargaming they chose their national color of Red as friendly[,] reversing the colors."

45 See McHugh, 2013.

46 U.S. Marine Corps 2005, quoted in W. Simpson, 2017.

47 McHugh, 2013, quoted in W. Simpson, 2017, who added the bracketed text. Note that Simpson attaches this definition to *umpire* rather than *controller*; his glossary contains entries for both.

48 Howard Key, "Expeditionary Warrior 2015 Wargame Facilitator Guide 150213," February 2015, quoted in W. Simpson, 2017.

higher authority typically operates outside the world of the game, but "some games may require the cell to be part of the game play instead of game control."[49]

Player: "A participant in a war game who is not a member of the control group, and who plays the role of a real-world commander of a staff officer of a military unit or units."[50]

Sponsor: "The senior officer or official of the command/organization who has requested the game. Normally, the sponsor is the approval authority on major game purpose, objectives, research questions, and design."[51]

Subject-matter experts (SMEs): "Participants who are experts on subjects related to the wargame."[52]

---

[49] W. Simpson, 2017.

[50] McHugh, 2013, quoted in W. Simpson, 2017.

[51] Burns, undated, quoted in W. Simpson, 2017.

[52] U.S. Marine Corps 2005, quoted in W. Simpson, 2017.

APPENDIX B

Defense Terminology Related to Information

In this appendix, we briefly review the lexicon related to IE itself.[1] We clarify several terms here to further differentiate their meaning and usage.

Information Environment

JP 3-13, *Information Operations*, defines the IE as "the aggregate of individuals, organizations, and systems that collect, process, disseminate, or act on information."[2] JP 3-13 further depicts the IE as three interrelated "dimensions" (see Figure B.1): physical, informational, and cognitive.[3] The cognitive dimension is at the top, perhaps implying a dependency on the lower two dimensions, informational and physical. In fact, the formal definition of the cognitive dimension states that it is the most important component of the IE:

> The cognitive dimension encompasses the minds of those who transmit, receive, and respond to or act on information. It refers to individuals' or groups' information processing, perception, judgment, and decision making. These elements are influenced by many factors, to include individual and cultural beliefs, norms, vulnerabilities, motivations, emotions, experiences, morals, education, mental health, identities, and ideologies. Defining these influencing factors in a given environment is critical for understanding how to best influence the mind of the decision maker and create the desired effects. As such, this dimension constitutes the most important component of the information environment.[4]

Interestingly, the definition of the physical dimension includes humans but not human decisions:

---

[1] This discussion is drawn almost entirely from Paul, Clarke, Triezenberg et al., 2018.

[2] JP 3-13, 2014, p. I-2.

[3] JP 3-13, 2014, p. I-2.

[4] JP 3-13, 2014, p. I-3.

Figure B.1
The IE as Conceptualized in JP 3-13

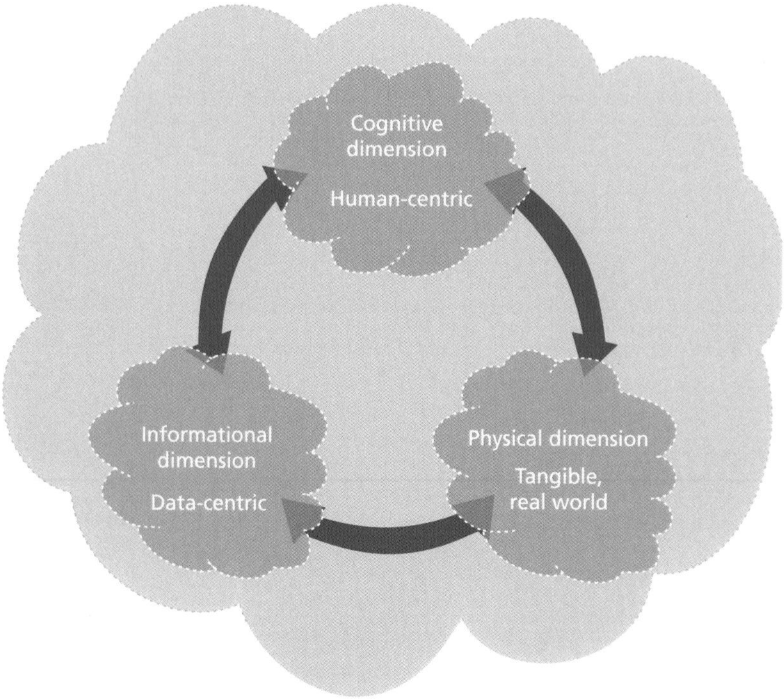

SOURCE: JP 3-13, 2014, p. I-2, Figure 1-1.

> The physical dimension includes, but is not limited to, human beings, C2 facilities, newspapers, books, microwave towers, computer processing units, laptops, smart phones, tablet computers, or any other objects that are subject to empirical measurement. The physical dimension is not confined solely to military or even nation-based systems and processes; it is a defused network connected across national, economic, and geographical boundaries.[5]

Finally, the definition of the informational dimension covers the means by which information flows, which can also be human-based.[6] The formal definition is as follows:

> The informational dimension encompasses where and how information is collected, processed, stored, disseminated, and protected. It is the dimension where

[5] JP 3-13, 2014, p. I-2.

[6] All three of the dimensions of the IE can be human-developed, but even the cognitive dimension could also include automated decisionmaking through artificial intelligence or other methods.

> the C2 of military forces is exercised and where the commander's intent is conveyed. Actions in this dimension affect the content and flow of information.[7]

When adding a "target audience" to the diagram in Figure B.1, JP 3-13 shows the human target as enclosed within the triangle formed by the cognitive, informational, and physical dimensions. As conceptualized in this manner, the IE is an environment that both shapes and confines the target audience. Both the IE and its dimensions are notionally depicted as clouds. The cloud symbol might communicate several characteristics of the IE: (1) that its boundaries are soft and, perhaps, changeable; (2) that its impact is largely cognitive (i.e., it concerns the realm of the mind); and (3) that it is ephemeral and has unclear physical boundaries.[8]

The DoD definition and understanding of the IE may be changing, however. At the time of this writing, JP 3-13 was under revision, and the October 2018 update to JP 3-0, *Joint Operations*, used a different definition of the IE:

> The information environment comprises and aggregates numerous social, cultural, cognitive, technical, and physical attributes that act upon and impact knowledge, understanding, beliefs, world views, and, ultimately, actions of an individual, group, system, community, or organization.[9]

The discussion continues, but it will continue to emphasize the human, physical, and informational aspects of the operating environment. Although the specifics of the language and the components emphasized appear to be changing, understanding of the underlying concepts appears to be consistent across these evolving definitions.

Information Operations

With the IE defined, the next term that likely comes to mind when considering military activities in and through the IE is *information operations.* The common-sense and colloquial understanding of IO takes the term at face value, assuming that IO are *operations* that have something to do with *information.* This understanding further suggests that IO personnel are operators who engage in these operations by employing information in some manner. This makes perfect sense, but it is not what *IO* is supposed to mean (and therein lies a possible problem).

IO, as formally described and practiced, are a planning, coordinating, and integrating function. In other words, it is a staff function, overseen by a staff officer, who

---

7 JP 3-13, 2014, p. I-3.

8 Rebecca Rosen has hypothesized that clouds "get traction as a metaphor because they are shape shifters, literally" (Rebecca Rosen, "Clouds: The Most Useful Metaphor of All Time?" *The Atlantic*, September 30, 2011).

9 JP 3-0, 2018, pp. IV-1–IV-2.

integrates the efforts of IRCs—efforts that are then executed by IRC personnel. These activities (or operations) should probably be called *IRC executions* or described by one of the other, newer terms in the lexicon. Unfortunately, such efforts have traditionally been colloquially mislabeled *IO*.

The relationship between the planning and integrating function known as IO and actual operations that use information is certainly similar to the relationship between *fire support coordination* and *fires*. Each pair of terms describes a staff function and a capability to execute or operate that function. But no trained member of the joint force would ever conflate fire support coordination with fires or expect a fire support coordination officer to leave a command post, travel to an artillery battery, and lay a gun. However, members of the joint force routinely conflate IO as a coordinating and integrating function with the execution of any effort in the IE (collectively incorrectly referred to as *IO*). They might well expect a staff officer whose task is planning and integration to go and lay out a storyboard for leaflets, get on a computer and do some cyber reconnaissance, or otherwise execute IRC tasks as part of operations (because, *operations*).

Not only does the current colloquial use of *IO* confuse the relationship between the planning and integration function and the actual execution of efforts in and through the IE, but the term is often used as shorthand for psychological operations/MISO.[10] This ignores the rest of the traditional IRCs and the inherent informational aspects of other military activities, including the presence, posture, and profile of deployed forces. Worse, a 2017 report by Facebook (which likely has a much larger readership than most DoD doctrinal publications) on false news and disinformation defined IO as "actions taken by organized actors (governments or non-state actors) to distort domestic or foreign political sentiment, most frequently to achieve a strategic and/or geopolitical outcome."[11] This definition promotes an understanding of IO that is inconsistent with both colloquial and formal DoD usage—and one that is quite pejorative. DoD would not want the joint force's use of the phrase *information operations* to invoke the Facebook report's definition for the wider public.

Because of these concerns, we limit our use of *IO* in this report and cite it only in its narrow, denotatively correct sense to describe a planning and integrating function.

Information-Related Capabilities

Misused less often than *IO*, but not wholly without contentiousness, is *information-related capability*. An IRC is doctrinally defined as "[a] tool, technique, or activity

---

[10] Curtis D. Boyd, "Army IO Is PSYOP: Influencing More with Less," *Military Review*, May–June 2007.

[11] Jen Weedon, William Nuland, and Alex Stamos, *Information Operations and Facebook*, v. 1.0, Menlo Park, Calif.: Facebook, April 27, 2017, p. 4.

employed within a dimension of the information environment that can be used to create effects and operationally desirable conditions."[12]

This is a perfectly reasonable and usable definition, except that it lacks clear boundaries. Almost anything that members of the joint force do or say can send a message or otherwise affect the IE, so almost anything could be an IRC, depending on the circumstances and consequences. Taken to its logical conclusion, such a wide-ranging definition, in practice, would require an enormous number of capabilities to be coordinated or deconflicted as IRCs.

While the joint definition is intentionally unbounded, certain capabilities are traditionally considered *information-related*. In fact, past IO doctrine listed five core capabilities alongside several supporting and related capabilities.[13] The traditional core capabilities were psychological operations/MISO, MILDEC, operations security, electronic warfare, and cyber operations. The supporting or related capabilities included public affairs, civil-military operations, defense support to public diplomacy, information assurance, physical security, physical attack, counterintelligence, and combat camera. Contemporary lists cite all of the above as IRCs. Service-specific materials have further listed key leader engagement and special technical operations as IRCs. Australian Army concepts have additionally included presence, posture, and profile in this category.[14]

Every action and utterance of the force can communicate a message or otherwise affect the IE, so we embrace the broad conception of IRCs and do not subscribe to specific or constrained lists.[15] Under this conception, some capabilities are *always* and *only* information-related, as their effects are limited to the IE. Other capabilities are *sometimes* or *secondarily* information-related; they are often used for other purposes and capable of having effects independent of the IE. Pretty much any DoD capability could be included in that second category, under certain circumstances. The *Joint Concept for Operating in the Information Environment* does not use the term *IRC*, but it *does* discuss both informational and physical power, and it emphasizes the importance of leveraging the inherent informational aspects of all military operations. In the context of bringing the IE into wargames, it is important to remember that effects in and through the IE come not only from IRCs but also from the echoes and implications in the IE of other military activities.

---

[12] JP 3-13, 2014, p. GL-3.

[13] See Christopher Paul, *Information Operations Doctrine and Practice: A Reference Handbook*, Westport, Conn.: Praeger, 2008.

[14] See James Nicholas, "Australia: Current Developments in Australian Army Information Operations," *IO Sphere*, Special Edition 2008.

[15] Christopher Paul, *Strategic Communication: Origins, Concepts, and Current Debates*, Santa Barbara, Calif.: Praeger, 2011.

As noted in Chapter Three, the term *IRC* may be falling out of favor. The latest revision of JP 3-0, *Joint Operations*, instead discusses "joint force capabilities, operations, and activities for leveraging information."[16] Under that heading, it lists the following functions (which would have traditionally been IRCs): key leader engagement, public affairs, civil-military operations, MILDEC, MISO, OPSEC, electronic warfare, combat camera, space operations, special technical operations, cyber operations, DoD information network operations, cyber-enabled activities, and commander's communication synchronization.

Operations in the IE

Relatively new on the lexical scene is the term *operations in the information environment*. The term was first embraced in DoD in the 2016 *Department of Defense Strategy for Operations in the Information Environment*.[17] A related term is embedded in the title of the *Joint Concept for Operating in the Information Environment*.[18] Interestingly, neither source defines *operations in the information environment* as its own term of art. Both define *IE* according to the doctrinal definition cited earlier in this appendix. Both then allow the standard definition of *operations* or *operating* to precede it, without any additional definitional discussion. We admire the implied simplicity: Once you have defined the IE, these are the operations that you undertake there.

IE Operations

The same words are used in a slightly different construction by the U.S. Marine Corps. A 2017 draft concept of employment defined *information environment operations* as

> [t]he integrated planning and employment of [Marine Air Ground Task Force], Naval, Joint, and Interagency information capabilities, resources, and activities that enhance the Marine Corps single-battle concept and provide defensive, offensive, exploitative effects and support in order to operate, fight and win in and through a contested information environment.[19]

The document that offered this definition explicitly distinguished it from IO, noting that these operations seek only cognitive advantage, while IE operations seek

---

[16] JP 3-0, 2018, p. III-22.

[17] U.S. Department of Defense, 2016.

[18] U.S. Joint Chiefs of Staff, 2018.

[19] U.S. Marine Corps, 2017b, p. 1. The single-battle concept emphasizes that a unified operational environment in which actions in one area can affect all parts of the environment.

any and all kinds of military advantage, including temporal, spatial, and technological. Marine Corps IE operations, then, are clearly envisioned as an umbrella concept that encompasses IO. The document further identified seven functions of IE operations, which are to be employed across six operational capability areas: electromagnetic spectrum, cyber, space, influence, MILDEC, and inform operations.[20] These operational capability areas appear to parallel the traditional core IRCs.

One Marine Corps stakeholder interviewed for this project informed us that, because the rest of DoD is employing *OIE* instead of *IE operations*, the Marine Corps would subsequently be calling its concept *OIE* (with the definition and treatment unchanged).

Information Warfare

Another term that appears occasionally in this context is *information warfare*. The term is not currently defined in joint or service doctrine, but it was in the 1990s. Chairman of the Joint Chiefs of Staff Instruction 3210.01 defined *information warfare* in 1996 as follows:

> Actions taken to achieve information superiority by affecting adversary information, information-based processes, information systems, and computer-based networks while defending one's own information, information-based processes, information systems and computer-based networks.[21]

The late Dan Kuehl of National Defense University defined *information warfare* more simply: "Military offensive and defensive actions to control/exploit the environment."[22]

Information warfare has come up in many recent discussions of military operations in the IE, even appearing in draft documents. In fact, the July 2017 *Marine Air Ground Task Force Information Environment Operations Concept of Employment* was, in draft form, titled *Information Warfare Concept of Employment* as late as May 2017, and

---

[20] U.S. Marine Corps, 2017b, p. 22. The seven functions are (1) assure enterprise C2 and critical systems; (2) provide IE battlespace awareness; (3) attack and exploit networks, systems, and information; (4) inform domestic and international audiences; (5) influence foreign target audiences; (6) deceive foreign target audiences; and (7) control information warfare capabilities, resources, and activities.

[21] Chairman of the Joint Chiefs of Staff Instruction 3201.01, *Joint Information Warfare Policy*, Washington, D.C., January 2, 1996.

[22] Dan Kuehl, National Defense University, "Information Warfare," briefing, undated.

the 2016 *Marine Corps Operating Concept* refers to information warfare and the integration of information warfare into the combined-arms approach.[23]

The term is attractive, as it clearly reflects a military or defense function. However, it also appears to confine these activities to "warfare." In fact, the joint force operates across the range of military operations—a spectrum that extends well outside of warfare to such activities as deterrence, shaping, humanitarian assistance, security cooperation, and conflict short of warfare. In an era partially defined by gray-zone aggression among competitors seeking gains in conflicts short of warfare, it would be prudent to avoid terms for operations in the IE that unnecessarily constrain the scope or range of those activities.[24]

Operations in and Through the IE

There are many terms that could be used to describe DoD plans and activities in and through the IE. Retired Army IO officer and longtime member of the DoD information community of interest Michael Williams has cautioned that "obsessing over the definition of information operations and what capabilities it may or may not include is a distraction."[25]

Whatever term ends up being embraced, it should have several characteristics. It should have all the usual virtues of a good definition. In particular, the common-sense interpretation of the term should correspond closely to the technical definition, and the term should not be in use with a different definition in a different community. It should also have some features specific to the requirements of defense efforts in the IE. The term and its definition should capture efforts and effects not only *in* or *on* the IE but also *through* it. The effects and efforts of greatest concern to DoD *transit* the IE to have an impact in the spatial domains. The term of art should also acknowledge that a broad range of activities have effects in and through the IE. Included capabilities should not be confined to those associated with messaging or with technical capabilities that affect C4ISR systems. Actions speak louder than words, and every action or utterance—or even the mere presence of the joint force—has potential echoes and consequences in and through the IE.

---

[23] U.S. Marine Corps, *Information Warfare Concept of Employment*, Washington, D.C., May 10, 2017a; U.S. Marine Corps, *Marine Corps Operating Concept: How an Expeditionary Force Operates in the 21st Century*, Washington, D.C., September 2016, p. 4.

[24] Christopher Paul, "Confessions of a Hybrid Warfare Skeptic," *Small Wars Journal*, May 3, 2016.

[25] Michael Williams, "Speed, Volume, and Ubiquity: Forget Information Operations and Focus on the Information Environment," *Strategy Bridge*, July 26, 2017.

APPENDIX C

Changes in the Marine Corps Related to Operations in the Information Environment

As noted in Chapter Two, The Marine Corps has created a Deputy Commandant for Information, as well as information groups in the Marine Expeditionary Forces.[1] The Marine Corps released its MAGTF IE operations concept of employment in July 2017, which lists the functions shown in Table C.1. It also includes a vision and framework for the 2025 time frame and outlines the organization and relationships for the new MIGs.[2]

The September 2017 Marine Corps Operating Concept emphasized the cognitive dimension of conflict, the importance of information as a weapon, the battle of signatures, future conditions under which "to be detected is to be targeted is to be killed," and the importance of information as part of combined arms.[3] The concept acknowledges that "the Marine Corps is currently not organized, trained, and equipped to meet the demands" of these IE requirements, so the service adjusted its force structure and introduced plans to develop new capabilities.[4] It established a three-star Deputy Commandant for Information, who converted the MEF headquarters groups into MIGs, changing their manning and composition for the new mission.[5]

These MIGs are a prominent organizational change intended to better align the operating forces with OIE. In June 2017, the Marine Corps appointed the first Deputy Commandant for Information.[6] Shortly afterward, in July 2017, the First Marine Expeditionary Force redesignated its headquarters group as a MIG. Although the headquarters group had been providing information capabilities, the intent was for the newly formed MIG to lead the planning, integration, and conduct information warfare with

[1] Pomerleau, 2017a.

[2] U.S. Marine Corps, 2017b, pp. 2–10.

[3] U.S. Marine Corps, 2016, p. 6.

[4] U.S. Marine Corps, 2016, p. 8.

[5] Pomerleau, 2017a.

[6] Mark Pomerleau, "Why the Marine Corps Needed a New Deputy Commandant," *C4ISRNET*, December 5, 2017b.

Table C.1
Seven Functions of MAGTF IE Operations

Function	Description
1. Assure enterprise C2 and critical functions	Actions to operate and defend networks, systems, and information to enable C2 and assured operation of critical systems
2. Provide IE battlespace awareness	Actions to characterize the physical, informational, and cognitive dimensions of the IE to identify challenges, opportunities, and comparative advantages of the MAGTF
3. Attack and exploit networks, systems, and information	Actions taken in accordance with approved authorities to exploit or attack adversary networks, systems, signatures, and information to create advantages for the MAGTF
4. Inform domestic and international audiences	Actions taken to inform domestic and international audiences to build understanding and support for operational and institutional objectives
5. Influence foreign target audiences	Actions taken in accordance with approved authorities to influence selected target audiences and affect their decisionmaking and behaviors to create conditions favorable to operational objectives
6. Deceive foreign target audiences	Actions to induce ambiguity, misunderstanding, resource misallocation, and delayed actions to mislead adversary decisionmakers and reveal their strengths, dispositions, and future intent while protecting the MAGTF's capability, readiness, posture, and intent
7. Control information warfare capabilities, resources, and activities	Actions taken to provide the commander with the ability to exercise C2; integrate assigned marine, naval, and joint information assets; and enhance the MAGTF's ability to operate in the IE

SOURCE: Lightly adapted from U.S. Marine Corps, Combat Development and Integration, 2017b, p. 2, Figure 1.

the goal of better integrating it into MAGTF operations.[7] Figure C.1 shows the MIG command structure as presented in the current concept of employment. The Marine Corps also identified several milestones for the MIG to achieve initial operating capability in fiscal year 2018.[8]

In addition to organizational changes, the Marine Corps has signaled interest in information in other ways. For example, a 2018 Commandant's Innovation Challenge, sponsored by the Marine Corps Rapid Capabilities Office, focused on eliciting ideas to man, train, or equip MAGTF IE operations.[9]

The establishment of the Deputy Commandant for Information and the release of a Marine Corps concept of employment for OIE were major steps reflecting the service's growing interest in and attention to the IE. The Marine Corps stood up

7 1st Intelligence Battalion, First Marine Expeditionary Force, homepage, undated.

8 "The Future Starts Now: Marine Corps Force 2025 Implementation and Information Warfare Capabilities," *Marine Corps Gazette*, August 2017, p. 79.

9 Marine Administrative Message 031/18, "2d Qtr FY-18 Commandant's Innovation Challenge," Washington, D.C., January 12, 2018.

Figure C.1
MIG Command Structure

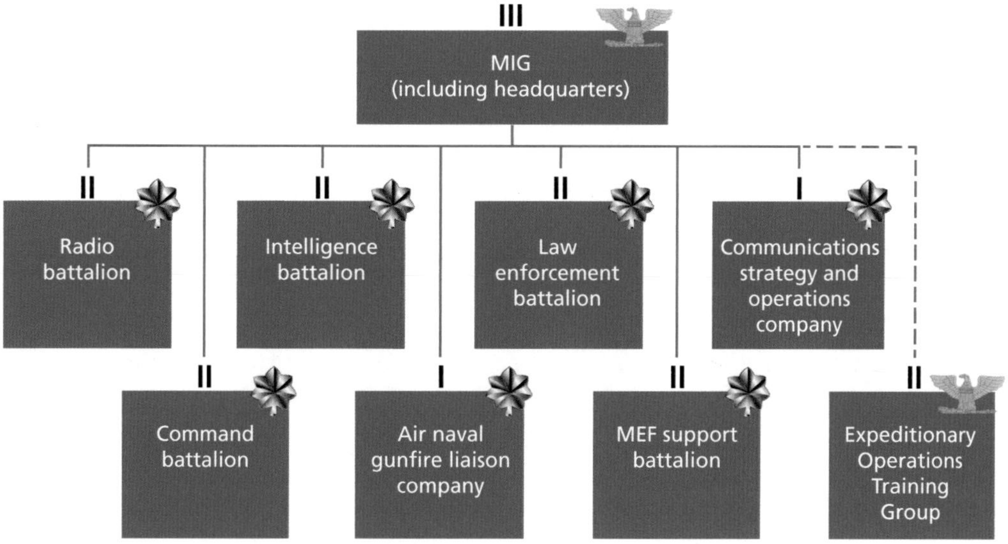

SOURCE: U.S. Marine Corps, 2017b, p. 4, Figure 2.

the Marine Corps Information Operations Center in 2009 to support the development of IO as a core military competency.[10] It subsequently established the Marine Corps Information Warfare Task Force in November 2015 and assigned an Assistant Deputy Commandant for Information Warfare to lead it. (After it was established, the office of the Deputy Commandant for Information absorbed these functions.) Initial task force responsibilities included identifying information warfare gaps, problem framing and mission analysis of information warfare in the future MAGTF, examining current capabilities that could contribute to addressing gaps, and developing conceptual and organizational constructs to integrate information warfare capabilities into the operating force.[11]

In line with information becoming a seventh joint warfighting function, the Marine Corps officially added information as a seventh Marine Corps warfighting function in January 2019. The Marine Corps bulletin establishing this function also directed the Deputy Commandant for Information to oversee its integration and for the

---

[10] Marine Administrative Message 0266/09, "Establishment of the Marine Corps Information Operations Center (MCIOC)," Washington, D.C., April 22, 2009.

[11] Marine Administrative Message 596/15, "Establishment of Marine Corps Information Warfare Task Force (MCIWTF)," Washington, D.C., November 25, 2015.

Deputy Commandant for Combat Development and Integration to update doctrine with the function description and school instruction with information terminology.[12]

However, the Marine Corps' increased emphasis on the IE in its doctrine, organization, and force structure has not yet permeated its wargaming apparatus.

Marine Corps Wargaming Practice Related to OIE

Current Marine Corps wargaming practice is largely consistent with that in the broader defense community, and it faces many of the same challenges when attempting to incorporate IE into wargames. For example, Marine Corps Title 10 games, such as the Expeditionary Warrior series, have historically been based on (usually kinetic) SSA scenarios and have used seminar-style formats.[13] Many other games and exercises, such as MEF exercises, have also traditionally focused on essential tasks supporting organizational training. Historically, these games have not treated information as a central concern.

Although the number of games devoted to OIE is growing, particularly under the new Deputy Commandant for Information, most games continue to follow a seminar-style format. Among the drawbacks of non-adjudicated seminar-style events is that they often do not challenge assumptions about the cause and effect or flow of operations. The Marine Corps Wargaming Division began offering adjudicated matrix games with SIGMAN 2018, but it still put reflected similar ideas about what might happen on both sides of the wargame.[14]

---

[12] Marine Corps Bulletin 5400, "Establishment of Information as the Seventh Marine Corps Warfighting Function," Washington, D.C., January 17, 2019.

[13] U.S. Marine Corps, *Expeditionary Warrior 2013: Future Maritime Operations for the 21st Century Operating Environment*, June 6, 2013, pp. 4–10.

[14] Discussion with Wargaming Division staff, Quantico, Va., July 12, 2018.

APPENDIX D

Wargaming Tools Reviewed

As part of our review of existing wargaming practices and tools, we examined the wargaming tools commonly used in the defense wargaming community.

Types of Wargaming Tools

Several types of tools are available to support wargaming. We considered tools that support knowledge management, visualization, adjudication, data capture, and analysis before, during, and after a game. We additionally considered such activities as information sharing and collaboration under knowledge management. Examples of knowledge-management tools for wargaming include chat software, online document-sharing tools, and online catalogs. Visualization tools include everything from online mapping software and PowerPoint to custom-built tools that can display the laydown of forces between wargame moves. Adjudication tools, intended to speed up the control cell's process for determining outcomes, are often purpose-built or highly specialized, because adjudication tends to be specific to an individual wargame system. On the other hand, analysis tools are often commercially available and include quantitative statistical packages, as well as qualitative data management and analysis software.

Although most professional wargames are still conducted manually, wargame planners sometimes employ computer-based tools designed for a specific process or context. However, such specialist tools are not particularly useful outside the specific purpose for which they are built.

State of Wargaming Tools for the IE

The wargaming tools commonly available for IE wargaming are limited, and only a few emerging tools appear able to handle the important IE elements and OIE dynamics that are missing from currently available tools. This gap reflects the fact that wargaming practices continue to focus on kinetic warfare, and learning from real-world operations in Iraq and Afghanistan has led games to adopt population-centric information efforts. These trends are compounded by the genuine difficulties of visualizing and tracking complex informational dynamics in a wargame and evolving concepts and understanding pertaining to the causal mechanisms behind OIE.

The Marine Corps also tends to have fewer technical tools than the other services. This may change in the future as it builds its wargaming capability, but, in the near term, Marines will likely continue to have limited access to certain wargaming tools. Given the relative immaturity of the tools available to wargame crucial IE dynamics, however, access will not be a limiting factor in improving most types of OIE wargaming.

Requirements for Wargaming Tools

In our review of wargaming tools, we investigated how these tools could support the requirements presented in Chapter Three and found that they could benefit wargaming in several ways:

- improved adjudication of IE effects
- qualitative analysis of player mental models and the effects of information on decisions
- better representation of C2 systems in games, including network and system-level effects
- improved understanding of cause and effect in C2 games
- realistic representation of cyber operations, electronic warfare, and spectrum management operations in wargames
- visualizations to support OIE.

Although a few *emerging* tools did appear to address the potential effects of degraded information in C2 games, we found significant gaps for the remaining requirements.

What might explain these gaps? In Chapter Three, we discussed a range of wargaming requirements for OIE. Wargame design typically begins with underlying theory for how one side or another prevails in military conflict. Next comes a discussion of defeat mechanisms, which are more concretely articulated and context-specific causal mechanisms for how underlying assumptions about the nature of conflict manifest. Wargames built to support various DoD activities include these prevailing assumptions and their associated defeat mechanisms in their mechanics. Wargaming tools support the game's mechanics and their development is often the last step in a game's design, because their use depends on these underlying theories of victory and posited defeat mechanisms.

However, what emerged from our review of the wargaming literature and practice were significant gaps in the assumptions, defeat mechanisms, and wargame mechanics appropriate for addressing OIE and the IE in general. For this reason, we expected to see few tools built to address information considerations in a way that is consistent with the view that information is central to the fight—and, indeed, this was the case.

Defense Wargaming Tools Examined

We reviewed the following tools to assess the state of the art in the defense wargaming community:

- Advanced Framework for Simulation, Integration, and Modeling (AFSIM)
- Analyst's Notebook
- Athena Socio-Cultural Model
- Atlas.ti
- Advanced Warfighting Simulation (AWARS)
- Applied Critical Thinking Handbook
- board and card simulations
- Close-Action Environment model
- CODE
- Combat XXI
- Command
- day-after methodology
- Decision Lens
- Decisive Action
- Dilemma analysis
- FacilitatePro
- Fleet Battle School
- future analytical science and technology (FAST) games
- Google Earth
- Green Country Model
- Harpoon
- hex games
- Joint Integrated Contingency Model (JICM)
- Joint Semi-Automated Forces (JSAF)
- Joint Seminar Wargaming Adjudication Tool 2 (jSWAT2)
- Joint Wargaming Analysis Model (JWAM)
- Map-Aware Non-Uniform Automata (MANA)
- map exercises
- massive multiplayer online wargames leveraging the internet
- matrix games
- modified commercial games
- Naval War College web apps
- RFLEX
- Simulation-Based Analysis and Training (SimBAT)
- simulation decks
- spreadsheet tools
- systemic operational designs
- standard wargaming integration and facilitation tools (SWIFT)
- ThinkTank
- versatile assessment simulation tools (VAST)
- virtual worlds

These are commonly used tools among defense wargamers across the services, the Joint Staff, Office of the Secretary of Defense, and Five Eye countries (Australia, Canada, New Zealand, the United Kingdom, and the United States). They run the gauntlet in terms of technological sophistication, from publications to computer simulations, and represent all the categories of wargaming tools discussed in this report. Excluded from this list are common software tools that wargamers use but that are standard and routinely available to nonwargamers. Examples of such excluded tools are Microsoft Office, SharePoint, and standard statistical analysis packages.

It is difficult to assess many of these tools for their applicability to OIE because their utility depends on how they are used in a wargame. For example, certain tools would be more accurately characterized as methods rather than tools; others are actually sets of tools or tools with multiple applications beyond wargaming.

Consider Day After methodologies, dilemma analyses, soft systems methodologies, applied critical thinking (formerly known as red teaming), and systemic operational design. In these cases, how OIE are represented depends to a great extent on how way the method is applied. Some of the tools we reviewed are commercial analytic tools used by the wargaming community, such as Atlas.ti, Analyst's Notebook, FacilitatePro, Decision Lens, and spreadsheet tools. Again, OIE- and IE-related content depends on the wargame's design. Others are classes of games, such as board and card games, hex games, matrix games, and map exercises, and the OIE content again entirely depends on what has been built into the game; they neither inherently contain nor exclude the IE. Others are primarily visualization tools, such as Google Earth, SWIFT, and VAST, display information-related content to the extent that this is a requirement of the game.

Defense simulations, such as AFSIM, AWARS, Combat XXI, and JSAF are powerful modeling and simulation tools with a heavily kinetic focus that were not necessarily designed with OIE in mind.

References

1st Intelligence Battalion, First Marine Expeditionary Force, homepage, undated. As of August 16, 2019:
https://www.imef.marines.mil/Units/I-MIG/1ST-INTEL-BN/Article/1253313/marine-corps-creates-first-information-group-to-prepare-for-modern-battlefield

Alkire, Brien, Sherrill Lingel, and Lawrence M. Hanser, *A Wargame Method for Assessing Risk and Resilience of Military Command-and-Control Organizations*, Santa Monica, Calif.: RAND Corporation, TL-291-AF, 2018. As of August 16, 2019:
https://www.rand.org/pubs/tools/TL291.html

Army Doctrine Publication 3-0, *Operations*, Washington, D.C., July 2019.

Army Field Manual 3-07, *Stability*, Washington, D.C., June 2, 2014.

Boyd, Curtis D., "Army IO Is PSYOP: Influencing More with Less," *Military Review*, May–June 2007, pp. 67–75.

Boyd, John, "Patterns of Conflict," briefing slides, December 1986. As of August 16, 2019:
http://www.ausairpower.net/JRB/poc.pdf

———, "Patterns of Conflict," video series with accompanying reconstructed slides posted by Dan Grazier, Project on Government Oversight, 2015. As of August 16, 2019:
https://www.youtube.com/playlist?list=PL4pmLxkc7CTcukIlpD0UThT7Y_K09oxXe

———, *A Discourse on Winning and Losing*, Grant T. Hammond, ed., Maxwell Air Force Base, Ala.: Air University Press, 2018. As of August 16, 2019:
https://www.airuniversity.af.edu/Portals/10/AUPress/Books/B_0151_Boyd_Discourse_Winning_Losing.pdf

Brightman, Hank J., and Melissa K. Dewey, "Trends in Modern War Gaming: The Art of Conversation," *Naval War College Review*, Vol. 64, No. 1, 2014, Article 4.

Burns, Shawn, ed., *War Gamers' Handbook: A Guide for Professional War Gamers*, Newport, R.I.: U.S. Naval War College, undated.

Caffrey, Matthew, Jr., "Glossary/Definitions of Wargame Related Acronyms and Words," Connections Wargaming, undated. As of August 16, 2019:
http://www.kentaurus.com/www.connections-wargaming.com/Definitions.html

———, lecture delivered as part of Air Force Materiel Command Wargame Course, c. 2016.

———, *On Wargaming: How Wargames Have Shaped History and How They May Shape the Future*, Newport, R.I.: U.S. Naval War College, 2019.

Chairman of the Joint Chiefs of Staff Instruction 3201.01, *Joint Information Warfare Policy*, Washington, D.C., January 2, 1996.

Cheng, Dean, "The People's Liberation Army on Wargaming," *War on the Rocks*, February 17, 2015. As of August 16, 2019:
https://warontherocks.com/2015/02/the-peoples-liberation-army-on-wargaming

Chivvis, Christopher S., *Understanding Russian 'Hybrid Warfare'—and What Can Be Done About It*, testimony before the Armed Services Committee, U.S. House of Representatives, CT-468, March 22, 2017. As of August 16, 2019:
https://www.rand.org/pubs/testimonies/CT468.html

Clark, James, "The U.S. Marine Commandant Wants a 'Star Trek'–Style Holodeck for Wargaming," *National Interest*, September 30, 2017. As of August 16, 2019:
https://nationalinterest.org/blog/the-buzz/the-us-marine-commandant-wants-star-trek-style-holodeck-22516

Compton, Jon, wargame designer, "Toward an Epistemology of Wargaming—A Drunkard's Walk," briefing, Military Operational Research Society, Wargaming Community of Practice Brown Bag Lecture Series, March 19, 2015.

Connable, Ben, Michael J. McNerney, William Marcellino, Aaron Frank, Henry Hargrove, Marek N. Posard, S. Rebecca Zimmerman, Natasha Lander, Jasen J. Castillo, and James Sladden, *Will to Fight: Analyzing, Modeling, and Simulating the Will to Fight of Military Units*, Santa Monica, Calif.: RAND Corporation, RR-2341-A, 2018. As of August 16, 2019:
https://www.rand.org/pubs/research_reports/RR2341.html

Curry, John, and Tim Price, *Matrix Games for Modern Wargaming: Developments in Professional and Educational Wargaming*, History of Wargaming Project, 2014.

Davis, Paul K., *Aggregation, Disaggregation, and the 3:1 Rules in Ground Combat*, Santa Monica, Calif.: RAND Corporation, MR-638-AF/A/OSD, 1995. As of August 16, 2019:
https://www.rand.org/pubs/monograph_reports/MR638.html

———, *Capabilities for Joint Analysis in the Department of Defense: Rethinking Support for Strategic Analysis*, Santa Monica, Calif.: RAND Corporation, RR-1469-OSD, 2016. As of August 16, 2019:
https://www.rand.org/pubs/research_reports/RR1469.html

Defense Modeling and Simulation Coordination Office, U.S. Department of Defense, *DoD Modeling and Simulation Glossary*, September 24, 2018. As of August 16, 2019:
https://www.msco.mil/MSReferences/Glossary/MSGlossary.aspx

DeLancey, Douglas J., *Adopting the Brigadier General (Retired) Huba Wass de Czege Model of Defeat Mechanisms Based on Historical Evidence and Current Need*, Fort Leavenworth, Kan.: School of Advanced Military Studies, U.S. Army Command and General Staff College, 2000. As of August 16, 2019:
https://apps.dtic.mil/dtic/tr/fulltext/u2/a393858.pdf

Dunford, Joseph F., Jr., "The Pace of Change," *Joint Force Quarterly*, Vol. 84, 1st Quarter 2017, pp. 2–3.

Dunnigan, James F., *Wargames Handbook: How to Play and Design Commercial and Professional Wargames*, 3rd ed., Lincoln, Neb.: iUnivers, 2000.

Dupuy, T. N, "A New Approach to a Breakpoint Methodology," draft article, February 19, 1987.

Echevarria, Antulio J. II, *Military Strategy: A Very Short Introduction*, Oxford, UK: Oxford University Press, 2017.

Fain, Janice B., Richard C. Anderson, Trevor N. Dupuy, Gay M. Hammerman, and Charles F. Hawkins, *Forced Changes of Combat Posture*, Fairfax, Va.: Historical Evaluation and Research Organization, Data Memory Systems Inc., September 30, 1988.

Friedman, B. A., *On Tactics: A Theory of Victory in Battle*, Annapolis, Md.: Naval Institute Press, 2017.

Fuller, John Frederick Charles, *The Foundations of the Science of War*, London: Hutchinson, 1926.

"The Future Starts Now: Marine Corps Force 2025 Implementation and Information Warfare Capabilities," *Marine Corps Gazette*, August 2017, pp. 77–80.

Gerwehr, Scott, and Russell W. Glenn, *The Art of Darkness: Deception and Urban Operations*, Santa Monica, Calif.: RAND Corporation, MR-1132-A, 2000. As of May 9, 2019: https://www.rand.org/pubs/monograph_reports/MR1132.html

Giles, Keir, *Handbook of Russian Information Warfare*, Rome: NATO Defense College, Research Division, Fellowship Monograph 9, November 2016. As of August 16, 2019: http://www.ndc.nato.int/news/news.php?icode=995

Heath, Timothy R., Kristen Gunness, and Cortez A. Cooper III, *The PLA and China's Rejuvenation: National Security and Military Strategies, Deterrence Concepts, and Combat Capabilities*, Santa Monica, Calif.: RAND Corporation, RR-1402-OSD, 2016. As of August 16, 2019: https://www.rand.org/pubs/research_reports/RR1402.html

Heath, Garrett, and Oleg Svet, "Better Wargaming Is Helping the US Military Navigate a Turbulent Era," *Defense One*, August 19, 2018. As of August 16, 2019: https://www.defenseone.com/ideas/2018/08/better-wargaming-helping-us-military-navigate-turbulent-era/150653

Hecht, Eado, "Defeat Mechanisms: The Rationale Behind the Strategy," *Infinity Journal*, Vol. 4, No. 2, Fall 2014, pp. 24–30.

Helmbold, Robert L., *Decision in Battle: Breakpoint Hypotheses and Engagement Termination Data*, Santa Monica, Calif.: RAND Corporation, R-772-PR, 1971. As of August 16, 2019: https://www.rand.org/pubs/reports/R0772.html

Helmus, Todd, Elizabeth Bodine-Baron, Andrew Radin, Madeline Magnuson, Joshua Mendelsohn, William Marcellino, Andriy Bega, and Zev Winkelman, *Russian Social Media Influence: Understanding Russian Propaganda in Eastern Europe*, Santa Monica, Calif.: RAND Corporation, RR-2237-OSD, 2018. As of August 16, 2019: https://www.rand.org/pubs/research_reports/RR2237.html

Joint Publication 1-02, *Department of Defense Dictionary of Military and Associated Terms*, Washington D.C.: U.S. Joint Chiefs of Staff, current as of July 2019. As of August 16, 2019: https://www.jcs.mil/Portals/36/Documents/Doctrine/pubs/dictionary.pdf

Joint Publication 3-0, *Joint Operations*, Washington D.C.: U.S. Joint Chiefs of Staff, incorporating change 1, October 22, 2018.

Joint Publication 3-13, *Information Operations*, Washington, D.C.: U.S. Joint Chiefs of Staff, incorporating change 1, November 20, 2014.

JP—*See* Joint Publication.

Key, Howard, "Expeditionary Warrior 2015 Wargame Facilitator Guide 150213," February 2015.

Kuehl, Dan, National Defense University, "Information Warfare," briefing, undated.

Leonhard, Robert R., *The Principles of War for the Information Age*, New York: Ballentine Books, 1998.

Linehan, Joey, "Expeditionary Advanced Base Operations (EABO) Wargame 2015 Design Brief," planning conference briefing, Quantico, Va., September 10, 2015.

Marine Administrative Message 031/18, "2d Qtr FY-18 Commandant's Innovation Challenge," Washington, D.C., January 12, 2018.

Marine Administrative Message 0266/09, "Establishment of the Marine Corps Information Operations Center (MCIOC)," Washington, D.C., April 22, 2009.

Marine Administrative Message 596/15, "Establishment of Marine Corps Information Warfare Task Force (MCIWTF)," Washington, D.C., November 25, 2015.

Marine Corps Bulletin 5400, "Establishment of Information as the Seventh Marine Corps Warfighting Function," Washington, D.C., January 17, 2019.

Marine Corps Doctrinal Publication 1, *Warfighting*, Washington, D.C., June 20, 1997.

Marine Corps Order 3070, *The Marine Corps Operations Security (Opsec) Program*, draft, June 1991.

Marine Corps Warfighting Publication 5-10, *Marine Corps Planning Process*, Washington, D.C., May 2, 2016.

Markley, James, *Strategic Wargaming Series Handbook*, Carlisle Barracks, Pa.: Center for Strategic Leadership and Development, U.S. Army War College, July 2015.

Mattis, James N., Secretary of Defense, "Remarks by Secretary Mattis on the National Defense Strategy," transcript, U.S. Department of Defense, January 19, 2018. As of August 16, 2019: https://dod.defense.gov/News/Transcripts/Transcript-View/Article/1420042/remarks-by-secretary-mattis-on-the-national-defense-strategy

Mattis, Peter, "China's 'Three Warfares' in Perspective," *War on the Rocks*, January 30, 2018. As of August 16, 2019: https://warontherocks.com/2018/01/chinas-three-warfares-perspective

Maybus, Ray, Secretary of the Navy, "Wargaming," memorandum, Washington, D.C., May 5, 2015. As of August 16, 2019: https://www.secnav.navy.mil/innovation/Documents/2015/05/WargamingMemo.PDF

McConnell, Dan J., *Investigating a Coherent Framework for Deception Analysis*, Monterey, Calif.: Naval Postgraduate School, June 1988.

McGrath, James R., "Twenty-First Century Information Warfare and the Third Offset Strategy," *Joint Force Quarterly*, No. 82, 3rd Quarter, 2016, pp. 16–23.

McHugh, Francis J., *U.S. Navy Fundamentals of War Gaming*, reprint, New York: Skyhorse Publishing, 2013.

McNerney, Michael J., Ben Connable, S. Rebecca Zimmerman, Natasha Lander, Marek N. Posard, Jasen J. Castillo, Dan Madden, Ilana Blum, Aaron Frank, Benjamin J. Fernandes, In Hyo Seol, Christopher Paul, and Andrew Parasiliti, *National Will to Fight: Why Some States Keep Fighting and Others Don't*, Santa Monica, Calif.: RAND Corporation, RR-2477-A, 2018. As of August 16, 2019: https://www.rand.org/pubs/research_reports/RR2477.html

Military Operations Research Society, "Certificate in Wargaming," registration page, undated a. As of August 16, 2019: https://www.mors.org/Events/Certificate-in-Wargaming

———, "Special Meetings," webpage, undated b. As of August 16, 2019: http://www.mors.org/Events/Special-Meetings

Murphy, Dennis M., *Talking the Talk: Why Warfighters Don't Understand Information Operations*, Carlisle Barracks, Pa.: Center for Strategic Leadership, U.S. Army War College, Issue Paper 4-09, May 2009.

Nash, John F., and R. M. Thrall, *Some War Games*, Santa Monica, Calif.: RAND Corporation, D-1379, 1952. As of August 16, 2019:
https://www.rand.org/pubs/documents/D1379.html

Nicholas, James, "Australia: Current Developments in Australian Army Information Operations," *IO Sphere*, Special Edition 2008.

Paul, Christopher, *Information Operations Doctrine and Practice: A Reference Handbook*, Westport, Conn.: Praeger, 2008.

———, *Strategic Communication: Origins, Concepts, and Current Debates*, Santa Barbara, Calif.: Praeger, 2011.

———, "The Crawl, Walk, Run Progression for the Integration and Conduct of Efforts to Inform, Influence, and Persuade," *IO Sphere*, Fall 2013, pp. 5–9.

———, "Confessions of a Hybrid Warfare Skeptic," *Small Wars Journal*, May 3, 2016. As of August 16, 2019:
https://smallwarsjournal.com/jrnl/art/confessions-of-a-hybrid-warfare-skeptic

Paul, Christopher, Colin P. Clarke, Michael Schwille, Jakub P. Hlávka, Michael A. Brown, Steven S. Davenport, Isaac R. Porche III, and Joel Harding, *Lessons from Others for Future U.S. Army Operations in and Through the Information Environment*, Santa Monica, Calif.: RAND Corporation, RR-1925/1-A, 2018. As of August 16, 2019:
https://www.rand.org/pubs/research_reports/RR1925z1.html

Paul, Christopher, Colin P. Clarke, Bonnie L. Triezenberg, David Manheim, and Bradley Wilson, *Improving C2 and Situational Awareness for Operations in and Through the Information Environment*, Santa Monica, Calif.: RAND Corporation, RR-2489-OSD, 2018. As of August 16, 2019:
https://www.rand.org/pubs/research_reports/RR2489.html

Paul, Christopher, and William Marcellino, *Dominating Duffer's Domain: Lessons for the U.S. Marine Corps Information Operations Practitioner*, Santa Monica, Calif.: RAND Corporation, RR-1166-1-OSD, 2017. As of August 16, 2019:
https://www.rand.org/pubs/research_reports/RR1166-1.html

Paul, Christopher, and Miriam Matthews, *The Russian "Firehose of Falsehood" Propaganda Model: Why it Might Work and Options to Counter It*, Santa Monica, Calif.: RAND Corporation, PE-198-OSD, 2016. As of August 16, 2019:
https://www.rand.org/pubs/perspectives/PE198.html

Perla, Peter P., *The Art of Wargaming: A Guide for Professionals and Hobbyists*, John Curry, ed., History of Wargaming Project, 2011.

Perla, Peter P., Mike C. Markowitz, Christopher A. Weuve, Stephen Downes-Martin, Michael Martin, and Paul V. Vebber, *Transforming Naval Wargaming: A Framework for Operational-Level Wargaming*, Alexandria, Va.: Center for Naval Analyses, September 2004. As of August 16, 2019:
https://www.cna.org/CNA_files/PDF/D0010807.A2.pdf

Pomerleau, Mark, "Marines Look to Dominate in Information Environment," *C4ISRNET*, April 5, 2017a. As of August 16, 2019:
https://www.c4isrnet.com/c2-comms/2017/04/05/marines-look-to-dominate-in-information-environment

———, "Why the Marine Corps Needed a New Deputy Commandant," *C4ISRNET*, December 5, 2017b. As of August 16, 2019:
https://www.c4isrnet.com/it-networks/2017/12/05/why-the-marine-corps-needed-a-new-deputy-commandant

Pournelle, Phillip, ed., *MORS Wargaming Special Meeting, October 2016: Final Report*, Alexandria, Va.: Military Operations Research Society, 2017. As of August 16, 2019: https://www.mors.org/Portals/23/Docs/Events/2016/Wargaming/MORS%20Wargaming%20Workshop%20Report.pdf?ver=2017-03-01-151418-980

Pournelle, Phillip, and Holly Deaton, eds., *MORS Wargaming III Special Meeting, 17–19 October 2017: Final Report*, Alexandria, Va.: Military Operations Research Society, April 2018. As of August 16, 2019: https://www.mors.org/Portals/23/Docs/Events/2017/Wargaming/MORS%20Wargaming%20III%20Report%20Final.pdf

Program Manager Wargaming Capability, Marine Corps Systems Command, "Wargaming Capabilities," Modern Day Marine 2018 Report to Industry briefing, September 26, 2018.

Rosen, Rebecca, "Clouds: The Most Useful Metaphor of All Time?" *The Atlantic*, September 30, 2011.

———, Director, Research and Analysis Division, Wargaming Department Center for Naval Warfare, U.S. Naval War College, "Using Wargames for Command and Control Experimentation," paper presented at the 8th International Command and Control Research and Technology Symposium, Washington, D.C., June 2003.

Selinger, Marc, "Air Force Eyes New Wargame on Multi-Domain Command and Control," *Defense Daily*, June 6, 2018. As of August 16, 2019: https://www.defensedaily.com/air-force-eyes-new-wargame-multi-domain-command-control/air-force

Simpson, Emile, *War from the Ground Up: Twenty-First-Century Combat as Politics*, Oxford, UK: Oxford University Press, 2012.

Simpson, William L., Jr., *A Compendium of Wargaming Terms*, last updated September 20, 2017. As of August 16, 2019: https://dnnlgwick.blob.core.windows.net/portals/0/NWCDepartments/Wargaming%20Department/A%20Compendium%20of%20Wargaming%20Terms%2020%20Sept%202017.pdf?sr=b&si=DNNFileManagerPolicy&sig=BEh3XmMzUbqnHm2SIms6QAWn5YWCMWBWnU8Ira1Oud8%3D

South, Todd, "Marine Wargaming Center Will Help Plan for Future Combat," *Marine Corps Times*, September 19, 2017. As of August 16, 2019: https://www.marinecorpstimes.com/news/your-marine-corps/2017/09/19/marine-wargaming-center-will-help-plan-for-future-combat

Storr, Jim, *The Human Face of War*, London: Continuum, 2009.

Tooke, Lamar L., "Blending Maneuver and Attrition," *Military Review*, Vol. 80, No. 2, March–April 2000, pp. 7–13.

UK Ministry of Defence, Development, Concepts, and Doctrine Centre, *Wargaming Handbook*, Swindon, UK, August 2017. As of August 16, 2019: https://assets.publishing.service.gov.uk/government/uploads/system/uploads/attachment_data/file/641040/doctrine_uk_wargaming_handbook.pdf

U.S. Department of Defense, *Department of Defense Strategy for Operations in the Information Environment*, Washington, D.C., June 2016.

U.S. Department of Defense Directive 8260.05, *Support for Strategic Analysis (SSA)*, Washington, D.C., July 7, 2011.

U.S. Department of State, International Security Advisory Board, *Report on Gray Zone Conflict*, Washington, D.C., January 3, 2017. As of August 16, 2019: https://2009-2017.state.gov/t/avc/isab/266650.htm

U.S. Joint Chiefs of Staff, *Joint Concept for Operating in the Information Environment (JCOIE)*, Washington, D.C., July 25, 2018.

U.S. Marine Corps, *Expeditionary Warrior 2013: Future Maritime Operations for the 21st Century Operating Environment*, June 6, 2013.

———, *The Marine Corps Operating Concept: How an Expeditionary Force Operates in the 21st Century*, Washington, D.C., September 2016. As of August 16, 2019: https://www.mcwl.marines.mil/Portals/34/Images/MarineCorpsOperatingConceptSept2016.pdf

———, *Information Warfare Concept of Employment*, Washington, D.C., May 10, 2017a.

———, *Marine Air Ground Task Force Information Environment Operations Concept of Employment*, Quantico, Va.: July 6, 2017b.

U.S. Marine Corps Warfighting Laboratory, Futures Directorate, "Title 10 Wargaming," webpage, undated. As of August 16, 2019: https://www.mcwl.marines.mil/Divisions/Wargaming/Title10.aspx

———, Wargaming Division, *Standard Operating Procedures (SOP)*, August 2005.

Utsumi, Takeshi, "Globally Collaborative Environmental Peace Gaming with Global University System," paper presented at the third International Learning GRID of Excellence Working Group workshop, Berlin, Germany, December 3, 2003.

Vebber, Paul W., Assistant Director, Wargaming and Future Warfare Research, Naval Sea Systems Command Warfare Center, Undersea Warfare Directorate, "Wargaming Compendium," email to the authors, July 8, 2015.

von Clausewitz, Carl, *On War*, Michael Howard and Peter Paret, eds., Princeton, N.J.: Princeton University Press, 1976.

Walton, Timothy A., *China's Three Warfares*, Herndon, Va.: Delex Systems, Special Report 3, January 18, 2012.

Waltzman, Rand, *The Weaponization of Information: The Need for Cognitive Security*, testimony before the Armed Services Committee, Subcommittee on Cybersecurity, U.S. Senate, CT-473, April 27, 2017. As of August 16, 2019: https://www.rand.org/pubs/testimonies/CT473.html

Watson, Bruce, *When Soldiers Quit: Studies in Military Disintegration*, Westport, Conn.: Praeger, 1997.

Weedon, Jen, William Nuland, and Alex Stamos, *Information Operations and Facebook*, v. 1.0, Menlo Park, Calif.: Facebook, April 27, 2017.

Whaley, Barton, *Stratagem: Deception and Surprise in War*, Cambridge, Mass.: Center for International Studies, Massachusetts Institute of Technology, 1969.

Williams, Michael, "Speed, Volume, and Ubiquity: Forget Information Operations and Focus on the Information Environment," *Strategy Bridge*, July 26, 2017. As of August 16, 2019: https://thestrategybridge.org/the-bridge/2017/7/26/speed-volume-and-ubiquity-forget-information-operations-focus-on-the-information-environment

Work, Bob, Deputy Secretary of Defense, "Wargaming and Innovation," memorandum, Washington, D.C., February 9, 2015.

Work, Bob, and Paul Selva, "Revitalizing Wargaming Is Necessary to Be Prepared for Future Wars," *War on the Rocks*, December 8, 2015. As of August 16, 2019: https://warontherocks.com/2015/12/revitalizing-wargaming-is-necessary-to-be-prepared-for-future-wars